George Weaver, James b......

Lectures on mental science according to the philosophy of

phrenology

George Weaver, James Burns

Lectures on mental science according to the philosophy of phrenology

ISBN/EAN: 9783741112706

Manufactured in Europe, USA, Canada, Australia, Japa

Cover: Foto ©berggeist007 / pixelio.de

Manufactured and distributed by brebook publishing software
(www.brebook.com)

George Weaver, James Burns

Lectures on mental science according to the philosophy of phrenology

LECTURES

ON

MENTAL SCIENCE.

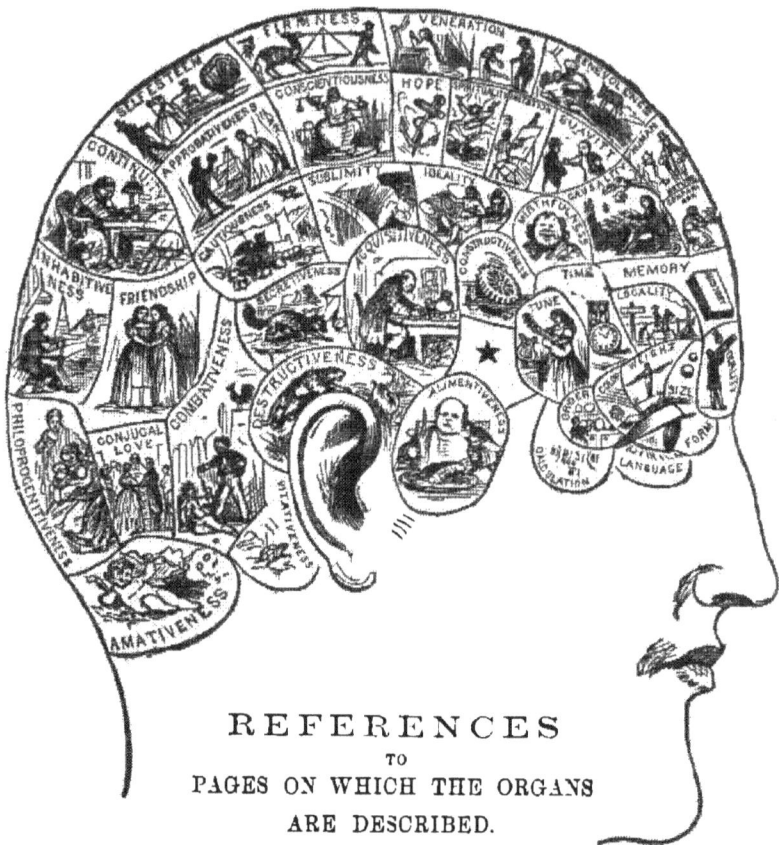

REFERENCES

TO
PAGES ON WHICH THE ORGANS
ARE DESCRIBED.

LECTURES

ON

MENTAL SCIENCE,

ACCORDING TO THE

Philosophy of Phrenology.

DELIVERED BEFORE THE ANTHROPOLOGICAL SOCIETY OF THE
WESTERN LIBERAL INSTITUTE OF MARIETTA, OHIO,
IN THE AUTUMN OF 1851,

BY THE REV. G. S. WEAVER.

NEW EDITION,
WITH SUPPLEMENTARY CHAPTER
LECTURER ON ANTHROPOLOGY.

LONDON:
JAMES BURNS, PROGRESSIVE LIBRARY,
15, SOUTHAMPTON ROW, W.C.
1876.

J. BURNS, PRINTER,
15, SOUTHAMPTON ROW, LONDON, W.C.

PREFACE.

THIS unpretending little work stands alone in the literature of the subject. It has been found that the more professionally-written works on Phrenology, overladen with technical detail, however valuable to the practical student, are dry and uninteresting to the general reader. This latter class is in all subjects by far the largest, and to them this volume is particularly dedicated. Not that the scientific matter presented is not as reliable as if it were clothed in language of a more professional character; for, while the pages bristle with anecdotes and popular allusions, and the style is rythmical as a poem, and entertaining as a tale, the view given of Phrenology is as correct and instructive, as far as space will permit, as that afforded by any other author. This is just the kind of book to induce that pleasing acquaintance with the science in its most applicable form, which, while it may lead to profound study of the subject, is eminently corrective to the reader in the most important issues of life. A high and holy religion, a pure spiritual philosophy, a liberal science, and elevating views of life in all its relations, are herein set forth.

The lectures were originally delivered in response to a request of the Anthropological Society of the Western Liberal Institute, Marietta, Ohio, in the autumn of 1851. The Society considered them " a valuable acquisition to Phrenological Literature," and " earnestly asked for the privilege of copying them for publication." In his reply, the author wrote :—

" These lectures have been thrown together at odd moments snatched from a multiplicity of arduous labours, and written at the electrical speed of the day. When essays are thrown into the printing-press at lightning velocity, who will be security for the reader while perusing them? Besides, what guard will there be against critics? Critics, you will reply, are harmless creatures; like barking dogs, they seldom do injury. True enough. And then, who writes for critics? Not the honest man, for he writes for *truth*. Not the good man, for he writes for the *good* of his readers. Not the brave man, for he writes in fearless determination of purpose. These lectures were written for the intellectual, moral, and social benefit of your society. If they have proved effectual to this end with you, they may with others. They have aimed at good. Their mark has been high. Their spirit is for progress. Their philosophy is the precept of the human soul's wisdom. Their morality is obedience to all divine law, written or unwritten. Their religion is the spirit-utterings of devout and faithful love. They aim at and contemplate humanity's good—the union of the human with the divine. The desire of your benefit, which alone prompted me to deliver them, now prompts me to comply with your request. Take them—transcribe them carefully—tell your printer and publisher to guard well against errors, and ask the world to read them in charity."

Having been valuable to many readers in the United States, the present edition is offered with the hope that these lectures may be of equal service to the young and inquiring mind of this country.

J. B.

15, SOUTHAMPTON ROW,
LONDON, 1876.

CONTENTS.

ANALYSIS OF THE HUMAN ORGANISM.

BY J. BURNS.

CLASSES OF ORGANS.		TEMPERAMENTS.		PAGE
PHYSICAL TEMPERA-MENTS.	VITAL APPARATUS.	1. Nutritive or Digestive	1	... 43
		2. Arterial or Active ...	2	... 44
		3. *Venous or Receptive*	3	... 37
		4. Pulmonary	4	... 44
		5. Excretory	5	... 44
		6. *Glandular or Lymphatic*	6	... 43
		7. *Cellular or Adipose*	7	... 43
	MECHANICAL APPARATUS.	1. Osseous or Bony ...	8	... 42
		2. Tendinous or Sinewy	9	... 42
		3. Muscular or Fleshy	10	... 7
PHYSICO-MENTAL TEMPERA-MENTS.	NERVOUS APPARATUS. (p. 7).	1. *Generative*	11	... 57
		2. Motive	12	... 13
		3. *Sensitive*	13	... 8
MENTAL TEMPERA-MENTS.	CEREBRAL APPARATUS. (pp. 9, 19—32, 45, & 49--55.)	1. Propelling	14	... 91
		2. *Restraining*	15	... 88
		3. *Domestic*	16	... 62
		4. Social	17	... 68
		5. Ipsial	18	... 80
		6. Perceptive	19	... 117
		7. *Recollective*	20	... 126
		8. Expressive	21	... 130
		9. Constructive	22	... 105
		10. *Æsthetic*	23	... 107
		11. Conceptive	24	... 127
		12. *Intuitional*	25	... 113
		13. *Moral*	26	... 132
		14. *Spiritual*	27	... 139
		15. Centripetal	28	... 146
		16. *Centrifugal*	29	... 135
		17. *Suggestive*	30	... 151
		18. *Critical*	31	... 149

[*This Diagram is Copyright.*]

NOTE.—The Temperaments printed in italics are those which constitute the negative or sensitive elements of character. The others are the positive or active. The former are spiritual and interior, the latter materialistic and demonstrative in manifestation. The groupings or temperaments may be somewhat explained by referring to the pages of this work indicated by the right hand column of figures, where the individual organs are approximately described. A general view of this system of classification is given in a lecture published in *The Medium*, No. 318, price one penny, and also in subsequent numbers of the same journal.

LECTURES ON MENTAL SCIENCE.

LECTURE I.

Physical Science mainly studied—Man known the last and least—The true
Mode not known—Man, the Ultimate of Creation—Mind studied in the
Abstract—Phrenology, the Key to Mind—How Man rules Creation—
Materialism—Error of Metaphysicians—Power of Mind over Matter—
Mind acting through Organism—Every Organ has a Specific Office—
Osseous—Digestive—Circulating—Nervous—The Mind alone Enjoys and
Suffers—Nervous Power and Sensibility—Large and Small Heads—Sus-
pension of the Mind—Injuries of Brain—Ignorance of Phrenological
Sceptics—Action of the Mind upon the Brain—Magnetically or Elec-
trically—God's Mode of Influence on the Universe—Phrenology the
Exponent of the Soul.

No truth is clearer than that "the proper study of mankind
is man." And yet how little is the real science of man studied!
It is the last and most neglected of all mental pursuits. The
physical and mathematical sciences have from time immemorial
occupied the attention of the men of genius and learning. The
earth has been circumnavigated again and again, its mountains
scaled, its bowels opened, its forests explored, its deserts tra-
versed, its jungles penetrated, its materials dissolved in the
crucible and separated by the blow-pipe, its agents and animals
classified ; the vault of heaven has been visited, its stars
counted and named, their velocities, magnitudes, densities,
orbits, revolutions, junctions, and appositions, and all their
grand and harmonious movements determined, to enrich and
perfect the physical sciences and to add wreaths of honour to
the tireless genius of man. In the great field of intellectual
labour, the busiest activity and the most vigilant energy and
perseverance have characterised the labourers. Names—proud

names—have been enrolled on the enduring scroll of fame, and
minds—rich, noble, powerful minds—have grown to giants, and
have weilded the mightiest sceptre of power over vast multitudes
of human beings, by the labours bestowed upon the sciences;
while, at the same time, *man* has been but slightly studied. A
passing thought only has been bestowed upon him. The
grandest, noblest subject of terrestrial observation, the lasting
pride, the quickening power, the fadeless honour, the crowning
glory of earth, has been passed by unstudied and unknown.
The grey-haired man of science has always gone down to the
grave, unconscious of what was within him, unknown to himself. The intricate machinery and nicely-adjusted systems of
his physical person, even, were hid from his view; while the
beautiful and majestic powers of his mental part were only contemplated with wonder and passed by as the " mystery of
mysteries."

Until very lately the field of mental science has been barred
against investigators. The key that would unlock the golden
treasure-house no man knew. Men have searched earnestly
for that key, that they might go in and labour diligently in
that field where the fruitage of heaven is growing on every
bough. They have made a thousand conjectures about what
was within. But to conjecture and speculation they have left
it all, feeling better satisfied to examine what the eyes could
see and the hands handle, than to press their inquiries in this
quarter. Nor was this unnatural. "First is the natural," or
physical, "afterward the spiritual."

It is expected that men will make themselves acquainted with
what they *see* before they enter the realm of the invisible. It
is expected that the more gross will antecede the more refined;
that the rudimental will precede the final; that the preparatory
will open the way for the graduating school. It is expected
that the child will be developed before the man; that the root
will strike downward before the stalk shoots upward to expand
its strength and beauty in the airs of heaven.

Man is the ultimate of earth, the last and noblest production
of creative skill connected with our planet, and, so far as we

know, with our universe. In him is centred the congregated perfections of all below, and the rudimental beatitudes of all above. He is the last link of the physical and the first of the spiritual, and hence we behold in him the reality of all that is earthly, and the promise of all that is heavenly—the animal and the angel—the evil and the good.

In his crimes and wickedness, he is the concentrated energy and essence of all animalty. In his wisdom and benevolence, he is a being of godlike attractions and powers, mighty in will, glorious in love. In him, earth and heaven meet; in him, matter and spirit unite. Spirit is active and powerful; matter is inert and impotent. Spirit rules; matter is subject. Spirit acts; matter yields. Spirit is the potter; matter is the clay. Hence, in this mysterious union, spirit operates upon, moulds, forms, animates, and in a great measure gives character to matter. Hence, spirit in man is seen, known, studied and judged of, through matter.

Spirit is the name given to free, untrammeled intelligence, to unfettered kindredness with God. Mind is the name given to spirit in matrimony with matter. Hence, mind is both kindred with God and brute, an inhabitant of earth, and a prospective emigrant to heaven. Being thus united in this inscrutable union, mind can be studied by mind only through the medium which makes it mind—through matter. This important truth has not been understood in years past. Hence, men have attempted to study mind as an abstract, indefinable something, separate and distinct from matter. They have rather attempted to study spirit without any means of acquaintance with it, upon merely abstract principles, as though it were possible to study spirit belonging to another sphere of being and action, another state of life and development, while belonging to this. These attempts, often exhibiting great strength of mind and loftiness of thought, gave good assurance that man was making diligent search for the long-desired key which should open to his admiring gaze and reflective genius the golden fields of mind waving with the ripened harvest of many centuries. That key was at length found in the discovery of Phrenology, the fundamental principle of which is based in the God-formed union of spirit with matter. Man had long studied matter. He knew its principal laws and arrangements, and hence from his previous advancement was well prepared to study mind through matter, or to study matter brought into immediate subjection or proximity to mind. That matter is in subjection to spirit, is the bottom principle of Phrenology. This ought not to be denied by any believer in spirit, by any believer in God. How moves God upon the countless myraids of material objects that throng the animated fields of His universe? How placed He them in their positions, gave them their harmonious movements, keeps them in their perfect order, sweeps the broad realms with the breath of His power, and glances through all the sparkling sunlight of His presence, if matter is not in abeyance to spirit? How is man lord of this lower creation; how does he sweep away the primeval forests that spread their giant arms above his home, and place in their stead his fields of fruit and grain; how ride securely on the bosom of the surging ocean; draw up the golden treasures of the earth and sea; make the rivers turn his million wheels, and snatch the lightnings from the clouds to become his pack-horse and mail-boy, if mind rules not the realm of matter? How account we for all we are and all we see; for the order, beauty, harmony, and magnificence of earth and sky, if above, below, beyond, in, and through all material forms, there dwells and rules not the omnipresent energy and intelligence of the Great Spirit? I repeat, then, the great truth, that spirit rules, forms, moulds matter, is the basilar principle of Phrenology.

Permit me here a momentary digression. It has been objected that Phrenology favours materialism, and hence joins hands with infidelity. So was Jesus accused of joining hands with Beelzebub. But let me ask, how looks this objection in the light of the principle I have just stated?

How spirit is joined with matter Phrenology professes not to say. That is knowledge, no doubt, that belongs to spirit, and not to mind. From questions which cannot be answered; from investigations which must from their nature be fruitless,

Phrenology turns willingly away, saying, "Let us labour and wait." It ventures not beyond the sphere of demonstrative reasoning.

The great error with past metaphysicians has been, in neglecting to acquaint themselves with the material connections of mind, and through these to seek an acquaintance with the principles of mentality. There can be no doubt that every exertion of the intellect, every flight of the imagination, every burst of passion, every glow of love, every feeling of sympathy, every emotion of joy or pleasure, calls into action some portion of the physical organism. A thousand daily phenomena gives us proof of this. A sudden fright, an outburst of passion, a rapturous joy, a burst of grief, any strong emotion, will give such a shock to the whole frame as to send the blood in leaping currents through every part and shake it like a trembling aspen from centre to extremity. Sometimes such sudden and strong exertions of mind have overpowered the physical frame, and caused it to dissolve in death. If such strong mental action produces such marked effect, then a feebler exertion of mind would produce a less effect upon the body. And so the conclusion follows, that every mental action produces a corresponding result upon the material organism with which it is connected. Why tires the body under mental exertion? Why shakes the frame in fear? Why blushes the face in shame? Why beams the countenance in joy? Why sparkles the eye in love? Why swells the bosom in grief? Why sickens the stomach in despondency? Why falters the tongue in embarrassment? Why softens the voice in sympathy? Why stretches the mouth in mirth? Why rolls the tear in affliction? Why beats the heart so wildly in any strong emotion? Who that denies that the mind manifests itself through the material organism, will explain all this? Why bows the *head* in sorrow? Why snaps it in anger? Why swings it in vanity? Why rises it high in dignity? Who will tell us that the mind manifests not its power and action through the body? But if the action of *any* portion or faculty of the mind affects the body, then the action of *every* portion of it does. And if a strong action of the mind makes an impression upon the physical substance with which it is mysteriously connected, then a weak one makes an impression also, only correspondingly weak.

The conclusion, then, is irresistible, that the mind does manifest its states and changes through the material organism with which it is united in a marriage of life. If, then, we can know the condition of the physical organism at any time, we can determine therefrom the condition of the mind. Hence, to study the mind we must study the physical organisation, for this is the medium, and the only medium, of mental manifestation. Through this, and only through this, can we trace the workings of the mind. In no other way do we, or can we, get any knowledge of it. He who attempts to study mental science, neglecting to attend to the physical, will fail, must fail; because this material structure is the only thing that has a positive union with mind. This, and this alone, opens the passage that leads to the sanctuary of thought and feeling. Here lies the mysterious pathway to the court of the soul. Without attending to this, all is conjecture, speculation, theoretical abstraction, doubtful ratiocination.

But here a question arises. Does the mind manifest itself through every portion of the body alike? or through some particular portion or member? So far as we are able to learn, each member of the body has a particular office to fill, and when the duties of that office are attended to, its work is done. Thus, the eye sees; the ear hears; the teeth masticate; the feet walk; the stomach digests; the glands secrete; the heart circulates; and so on to the end. They each have the duties of one office to perform, and only one. Can we suppose that they must all take on the extra and arduous duty of being the medium of mental manifestation? This is an unreasonable supposition when we have no proof of it. The truth is, each organ of the body has its single and particular function, and this only.

But the body has several great systems; may the mind not be manifested through one or all of these?

The *osseous* system is the framework of the body, and is composed of hard, mineral bones, so morticed, grooved, and bound together as to form the skeleton, to hold the different parts of the body in their places, and give it strength and locomotion. The function of this system is plain ; so it cannot be to manifest mind.

The *digestive* economy, consisting of the stomach and bowels, and the glands of the chest and viscera, has its function most clearly defined. It is to digest and prepare nourishment for the body.

The *circulating* system, consisting of the heart, lungs, arteries and veins, is to oxygenate, electrify, and warm the entire system.

The *muscular* system is to bind, strengthen, and beautify the whole, and give it its power and ability for locomotion. All these systems have their distinct uses.

Now comes another, the *nervous* system, the most intricate and delicate of all ; but little known until lately ; running through and ramifying every portion of the whole body ; endowed with the highest possible degree of delicacy and sensitiveness. What is it for? It is of such a nature that it can perform none of the offices of the others. It is securely guarded, placed in the deepest and most secure positions in the body. Its fibres or threads are composed of a soft, white, or greyish-white substance, exhibiting scarcely anything like texture, and all connected in one million-threaded web, or tree, having its base in the brain, which is the nucleus or grand centre of the whole system. Touch but the least possible one of these nerves, though with the point of a cambric needle, and the whole system is thrown into convulsions with the quickness of thought. All sensation, feeling, pain, and motion, are confined to this system ; but not the function of life ; for sever a nerve, as the nerve that leads to a limb, and the limb will live on as ever, but destitute of all feeling. That all sensation, pain, and pleasure is effected through the medium of the nerves, is now a point settled beyond controversy. But what is sensation? Is it a bodily, or a mental feeling? Does the body, of itself a congregate mass of clay, possess sensation? Is it possible?

Whatever possesses sensation doubtless feels pleasure and

pain. I know it is a common form of speech to speak of "bodily pains and pleasures." But can the body feel pain or pleasure? Can clay suffer or enjoy? The conclusion is unavoidable, if this subject is examined, that it cannot. If, then, no part of the body can suffer, or enjoy, or possess sensation, what does possess these susceptibilities? The only answer is, the mind. The mind alone suffers, enjoys, knows, appreciates, takes cognizance of motion, size, form, colour, and all those properties which are addressed to the senses. But we have seen that the nervous system is the medium through which these effects are produced. The conclusion, then, seems pressing upon us, that the nervous system is the mental medium—the medium through which mind acts, the servant which it employs to communicate with the outward world. It is through the senses—seeing, hearing, tasting, smelling, and feeling—that we get our knowledge of the outward world. The effects which outward things produce on the body we know are recognised by the nerves. But as the nerves are clay, material substance, and cannot of themselves possess sensation, we must conclude that they are employed by that in which resides all power of feeling, thought, and action—by mind—for the purpose of connection with the outward world. The brain, we have stated, is the grand centre of the nervous system. Every nerve of the body is connected with it. The brain, as it were, forms the great root or base of the whole system. The nerves of the head and face connect immediately with the brain. But the great majority of the nerves connect with the great spinal nerve, which forms a sort of trunk, growing out of the brain, and extending down the vertebræ of the back. From this, numerous branches of nerves lead off to every part of the body, ramifying it with a million thread-like divisions. Thus does the whole system join with and centre in the brain. From this arrangement, it is evident that the brain is the most important part of the nervous system. Here is the centre, the power, the life of all. What the heart is to the circulating system, the brain is to the nervous.

But here comes the most important question in physiology: What is the real use of the brain? There are many objectors to Phrenology; but they have never told us the use of the brain. Read the hundreds of physiological writers of the past, and you will nowhere find an office given to the brain equal to its manifest importance in the human system. It stands, as it were, the crown of the whole body, erected upon the highest point, guarded in a most wonderful manner, composing the great bulk of the nervous material, supplied with one fifth of the blood of the whole body, using one fifth of the nourishment taken into the body; showing that its labours are great, and its office a paramount one, in the highest degree sensitive, and from all these considerations, evidently in the highest degree useful. Now, who among all physiologists before Gaul and Spurzheim—who among all opposed to, or unacquainted with, the science of Phrenology, has told us what office the brain performs equal to its size, position, and call for nourishment? I unhesitatingly answer, *no one*. The brain has evidently been a great mystery to the physiological world. Phrenologists assert that is the dwelling-place of mind, the grand throne-room of spirit, the great machine-shop of the soul, from which is sent out the thousand inventions, reports, sciences, speeches, dramas, poems, books, monuments of art and wisdom, which have marked the career of man, and wreathed his brow with imperishable honours. The proof of this is founded upon experiment and observation, upon *facts* which are daily staring us in the face.

First of all is the fact, that man is lord of this lower creation.

He rules with despotic sway over the countless tribes of huge animals that throng the forests and rove the plains of his terrestrial home. They are his superiors in strength, size, and ferocity. A single one could tear a thousand men into atoms in a trial of physical strength. And yet he is lord, proud and mighty in conscious strength and authority. He rules them by the power of his mind. They are his physical superiors in every thing but *brain*. In this particular he towers a world above them, as much as he does in his mental strength. Look at man. See the massive lobe of brain that rises above his eyes and ears; a dome of strength, an arch of grandeur. Compare

this elevation with the upper-head of any animal. The contrast
is as wonderful as is the contrast of mind between the two.
The animal economy is carried on as perfectly in animals as in
man. They are as healthy, as active, as strong, and exercise
the various senses as actively as man. This proves that this
vast lobe of brain is not necessary for any office, and is not
called to perform any function in the animal or organic economy.
The body will perform all its offices and functions just as well
with a spoonful of brain, or with enough to form a nervous
centre, as with the great measure-full that man has to carry
about. It is perfectly evident that this great quantity of brain
is not for any use in the merely animal or organic economy.
Then what is it for? Pray, who will tell us, if it is not for
the uses of mind? Prove to us that it has any other use in
the human economy equal to its importance and position, and
we, phrenologists, will give up our theory, and push our researches in the science of mind in some other direction. Come
forward, objectors; come, all opposers of Phrenology, and tell
us why man is burdened with such a load of brain—why is
piled up away on the top of his system this huge skull-full of
clay, made so very nicely that he must take the very best
possible care of it, or he will become a simpleton, or a fool, or
a madman, or lose his life, or some other awful thing? Was
this one of the curses of the "*fall*" with which man was
loaded? Was this the wretched freight that the poor pilgrim
had to carry about, which Bunyan has so graphically described?
Suppose we clip it off. Try it, Mr. Objector; amputate it.
You would not, of course, at all affect the *mind* by it. No, no.
The brain has nothing to do with the mind, no more than the
foot or the hands. Then take it off down close to the eyes, and
see how much *mind* you have left. See how much your brain-
less man would be above an animal. There is a strong pre-
sumptive argument in this view of the subject in favour of the
phrenological position, that the brain is the medium of mental
manifestation.

Another fact bearing upon this point is, that strong minds are
generally connected with large and active brains. If there are

exceptions to the truth of this remark, they are explained on the ground of intense activity. Look at the heads of our philosophers, statesmen, men of genius, men who have moved the world, to whom multitudes have listened with breathless attention, and whom nations have praised and half deified. They are large and high. They have enormous brains. And if this is the curse, then the best men are cursed most heavily ; while all natural fools and very weak-minded persons, unless their idiocy is occasioned by disease or accident, or unless they have a stupid, sleepy, almost lifeless system, possess little, cramped, lilliputian heads.

Another fact, and a very stubborn one, is, that there is as great a diversity to the form of the human brain as there is to the human character, and a close correspondence between the two ; so that no man of a low, flat top-head, ever possesses an elevated moral character ; or of a narrow, low, cramped, short forehead, ever possesses a strong logical, philosophical intellect ; or of a small, flat back-head, ever possesses strong and confiding social affections.

Still another fact is strong upon this point. Diseases and injuries of the brain, pressure upon it, &c., will always derange, or entirely suspend, the mental operations. Take a healthy, sound man, remove a portion of the skull, then with your finger press upon the brain, and all consciousness will be suspended, all mental power, all feeling, so that you can cut his body in pieces, and he will not know it. Remove your finger, and instantly his mental consciousness, power, and sensation will return.

These facts, with numerous others which phrenologists have observed, during long years of patient study and investigation in the dissecting-room, the insane asylum, the hospital, and the grand theatre of the world, in the examination of millions of heads and a comparison of them with their known characters, have established in the minds of candid men the position, that the brain is the medium of mental manifestation. Those who have denied this position, have done it without examination. In every instance, as far as my knowledge extends, the denial has been made in ignorance. Let them go into the field, and prove their denial by actual demonstration, and then it will be of some value. Till then it will be regarded by every phrenologist as the croaking "see-saw" of ignorance.

How spirit, or mind, makes use of brain in manifesting its powers, Phrenology pretends not to say. This is a question which does not come legitimately within the sphere of phrenological inquiry, and probably is a question which is not capable of a positive answer at present. The *how* of anything is always the last to be reached.

It may not be improper for me to venture a few suggestions, which have pressed themselves upon my own mind with much force, touching the question: How does mind act upon brain in its manifestations? From numerous experiments and observations, it has been learned that the nervous system, or the nerves, are most perfect conductors of electricity, and that when the nerves of the body of a man or animal, just robbed of life, are touched with the charged wire of a galvanic battery, or an electrical machine, it will immediately exhibit the most striking and often frightful symptoms of life, by leaping, twitching, writhing, and hideous contortions of countenance. From many similar experiments, it is made more than probable that the most important function of animal life is performed by electrical agency. Pressing these electrical observations still further, under the experiments of "animal magnetism," the conclusion seems more than probable, that the brain is a most perfect galvanic battery, generating perpetually, in a state of health, a constant flow of electric fluid, and connecting with every part of the system by means of the nerves. These nerves are mediums of communication, by electrical agency, between the brain and the outer world. Electricity is the most subtle, ethereal, all-pervading agent of which we have any knowledge, and is the best adapted to perform the offices of mind with the outward world, of any known agent. Taking the facts that the brain is a galvanic battery and is also the medium of mental communication, the conclusion is very plausible that the immediate agent in this communication is electricity. Perhaps not in the form in which it acts to produce the effects which we see connected with material substance; it may be in a more refined, ethereal form. But there is great reason to believe that electricity, in its nearest approximate to spirit, in its most refined and powerful state, is the agent of mental communication. The mind makes use of this refined and almost spiritualised agent, generated by the brain, to convey its thoughts, states, moods, and feelings to the world. If this view is correct, then the brain is a most finished and complicated telegraphic office, connected with all the outward senses and every part of the body by the nerves, which are really telegraphic wires, conveying intelligence from the brain outward, and from the outward senses inward to the brain. The mind is the telegraphic officer who gives and receives despatches. Whenever any impression is made on any of the outward senses—the eye, ear, olfactory, gustatory, or nerves of feeling—whether it be pleasant or unpleasant, of heat or cold, of beauty or deformity, of pain or pleasure, a report of that impression is carried from the place where it is made, by the nerve which connects with it, and is read in an instant in the great central office, where the officer is always in waiting, attending to the calls that come in from every part. The decisions of that officer, or his determinations, are carried back by another set of nerves, the moment he feels disposed to return them. There are two sets of nerves, one called the nerves of motion, the other the nerves of sensation. The nerves of sensation communicate from without to the mind within, and the nerves of motion from the mind to the various parts of the body, commanding them to perform the dictates of the will. Thus : that book lies in my hand. The nerves from the hand, and also from the eye, report to the mind in an instant, "A book in hand." This report is carried in by the nerves of sensation. The mind replies immediately, " Open and read." It sends this report out on the nerves of motion, and immediately the hands move as directed, the eye turns towards the open pages, and commences the work of reading. The nerves of sensation carry back to the mind an account of the letters, words, thoughts, &c., found on the page. When

the first page is read, the nerves of sensation announce the
fact. Immediately the mind replies, through the nerves of
motion, "Turn over," and the hands perform at once the proper
motions to turn the leaf over. In this manner, all outward
sensations, all pleasure and pain, are felt by the mind; and all
motions, actions of the body, limbs, members, &c., are directed
by the mind. Thus the brain becomes the grand instrument
in performing the varied and multiform actions of the living,
acting man, while the subtle and mysterious fluid, quick as
thought, and ethereal almost as spirit, which the brain collects
and holds, is made the immediate agent of mind in performing
its wonderful evolutions. It is in this way, no doubt, that
mind operates upon matter, and rules and moulds it to its will.
It is very likely that by the use of this same all-pervading, all-
powerful, and invisible fluid, the mind, or spirit of God moves
upon, moulds, and controls the grand movements of His
illimitable universe. This is His waiting servant, standing by
the throne day and night, which spreads out its resistless
influence from the centre to the circumference of creation. It
is not impossible that God has a particular home in the heaven
of heavens, where He dwells "*in propria persona*," the central,
all-holy of holies, the throne-room and presence-hall of creation's
august Monarch, which is the centre of all electrical influence,
into which is poured the momentary reports of all worlds and
all creatures; while from it is issued, in ceaseless wisdom and
love, the mandates of Almighty power, darting through the
universe with the commissioned energy of God's omnific
presence. It is not impossible that by this means God makes
Himself omnipotent and omnipresent. Who can say that this
is not the agent by which God rules in the armies of heaven,
and among the inhabitants of earth? So far as we know anything of His grand course of procedure, He uses means in the
accomplishment of His ends. The mightiest, subtlest, most
universal and powerful agent in our world known to man, is the
mysterious and invisible one of which we are speaking. It is,
doubtless, the secret spring of all motion, mutation, life, and
death in the animal kingdoms. The laws that govern our world,

I doubt not, govern all worlds. What electricity is here, it is everywhere.

May not, then, electricity be the universal agent by which mind rules over matter, whether the mind be finite or infinite? These speculations belong not legitimately to Phrenology. I have thrown them out while speaking of the uses of the brain, because to me they are highly charged with probability. Phrenology stands on positive ground. Its asks nothing only what it can *prove*. It yields nothing only what is proved. Theories it gives, like chaff, to the wind. One fact it regards as worth a million of them. It treads on facts at every step. It is the product of experiment. It is eminently the science of daylight. It has come out from under the hand of the dissector and the manipulator.

It is a mental and moral science. It proposes first to teach a man himself—the most important knowledge within his reach, and the one in which most men are most unaccountably deficient. It would unfold the closely-drawn curtains of self. It would map out a chart of the soul. It would expose the motive springs of all actions, tell a man why and how he feels and acts. It would open the sweet-scented garden of the affections, and count and name each flower of love, and tell its peculiar fragrance. In a word, it would picture a man's soul on canvas and hold it up for him to look at just as it is, with its beauties and deformities strangely congregated. It would then point out its faults, its, weaknesses, its dangers, its darling propensities; then tell him how to improve it, how to curb its passions and quicken its aspirations, how to refine its coarseness and empower its energies, how to correct its judgment, enlighten its reason, purify its love, elevate its sentiments, beautify, adorn, and perfect its character. In a word, it would give him that knowledge by which he could harmonise himself, form a perfect mind within him—the most beautiful, grand, glorious, sublime thing in earth, that which angels admire in rapture, and God Himself loves in infinite ardour. It would confer upon every man, every woman, the priceless boon of this knowledge. It would join and cement for ever the links of golden friendship;

consummate the nuptial bonds of congenial spirits; open to their enraptured eyes the pure, refined, and ecstatic pleasure of a love such as binds perfected souls in a bliss that knows no end; and lay in them the deep and sure foundation for a higher, purer, nobler, truer race of men and women, with which to construct the sublimely noble fabric of a peaceful, harmonious, religious, enlightened, and happy society. With this for its object Phrenology goes forth; and may the winds bid it speed, the waters bear it on, the lightnings write its message on the heavens above, and the hearts of men clap their hands in jubilant joy wherever it goes !

LECTURE II.

Do different Parts of the Brain manifest different Faculties ?—"The Brain a Unit," ridiculous—Several Faculties act at once—Insanity of single Faculties—Form of Brain indicates Character—Size of Brain as Strength of Mind—Texture a Measure of Power—Balance and Activity of Brain—Adams and Webster contrasted—The Skull shows the Form of the Brain —How the Brain expands the Skull—Structure of the Skull—Form of Head shows Form of Brain—The Outward Man the Voice of the Inward—Health of Mind as Health of Body—Natural Language of the Organs—Power of the Actor, Orator, and Poet—Description of the Brain—The Presence-Chamber of the Mind—Convolutions of the Brain—Exercise increases Size and Power—Harmony and Balance of Mind.

WE closed our first lecture with the consideration of the brain as the medium of mental communication. It may not be improper to introduce the present with the question : Does the whole brain act in every effort of each of the several faculties of the mind? Does every mental act or emotion call into exercise the whole brain ? On this it is not irrelevant to assert, what the whole metaphysical world has long since, as by common consent, acknowledged as truth, that the mind is composed of several distinct faculties, as reason, imagination, affection, pride, anger, &c.

Now, so far as we know anything of the nervous system, it works upon the plan of an apportionment of labour. To each part is given a particular labour. The nerves of the eye are

for sight; those of the ear are for hearing; those of the nose
for smelling, &c., each being set apart to a distinct work. Now,
as the different faculties of the mind are very different in their
natures, we should suppose that they would require different
organs for their manifestations. As loving and reasoning are
so widely different in their characters, it is but reasonable to
suppose that they would manifest their powers through different
and widely-separated organs. Reasoning from the analogies
furnished us by every other portion of the system, we must
conclude that the whole brain is not used in every minute act,
but that a particular portion only is made requisite. Again, if
the whole brain is used in each mental act, it must, it would
seem, get dizzy sometimes in its multiform and rapid changes.
Sometimes it reasons, sings, prays, persuades, fights, and loves,
in about as short a space of time as it has taken me to read
this sentence. And these may be followed by laughing, crying,
deceiving, hoping, and fearing, in the next half minute; and
these followed again by other states of mind in rapid succession.
Now, if the whole brain is made to serve thus rapidly each
one of these mental powers and exhibit their different phases,
burning with the hot flames of so many passions in such quick
succession, it has a task more laborious than that assigned to
Hercules. How it could be the servant of so many stormy
masters in so short a time is not easily conceived.

Again, two or more mental faculties may be acting at the
same time. For instance, I may be carrying on this course of
reasoning, and watching the nods of some of my sleepy hearers,
and be deeply mortified at their want of attention. One may
be thoroughly angry with an enemy, and, at the same time,
use his reasoning powers to lay plans to effect the destruction
of that enemy. One may be deeply in love with an object of
great interest, and, at the same time, employ his imaginative
intellect to write poems in praise of that object. You may be
listening to me and catch every thought of my discourse, while,
at the same time, you may be in a state of despondency or of
real joy, at the thoughts which you are entertaining of some
beloved one in the distance. Or you may be, while I am

proceeding, applying my thoughts to other subjects and other uses entirely foreign to this lecture. Indeed, if you will examine carefully your own mental states, you will find that they are more often double than single. How, then, can the whole brain, a distinct unity, be used to manifest two or more distinct mental states at the same time, or be made the servant of many separate mental powers at once? See here—how do I coin my thoughts and select the proper language with which to express them, if I use the whole brain both for coining my thoughts and choosing my language; for I arrange my words while I am forming my thoughts? These considerations are directly at war with the idea that the whole brain is used in every mental process.

Permit another thought upon this point. It is not unfrequent that cases of insanity occur, in which the mind remains sound except in a single faculty. Sometimes it is the religious, sometimes the affectionate, sometimes other portions of the mental constitution that are thus affected. How can this be explained if the whole brain be used in every mental act?

The phrenological conclusion is, that every mental power is manifested through a single cerebral organ, or a particular portion of the brain, which is devoted exclusively to this work, just as each nerve or member of the whole body performs a single office. In proof of this, besides the considerations I have mentioned, there are many others which have forced themselves upon the attention of the practical phrenologist.

In cases of diseases of the brain, it is found that diseases and injuries in a certain portion of the brain always affect a certain portion of the mind—derange a certain mental faculty. It is thus that partial insanity is explained and reduced to philosophical principles.

Again, the long-continued and powerful exertion of any mental faculty always brings on a cerebral derangement in a particular portion of the brain, so that partial insanity is produced by this undue mental action. Observations of this kind have shown that insanity belongs not to the mind, but to its medium of manifestation. Cure the medium, and the mind is always cured. Once more, it is found that when any particular mental faculty is strong, a cerebral development, correspondingly strong, always accompanies it. On this one fact hang the most overwhelming proofs of the truth of phrenology, and as the size and conformation of the brain may be known by the size and outward form of the head, the science may be tested by every one who has hands or eyes and a tolerable degree of common sense.

The sum is this; the form of the head is the outward evidence of the character of the mind, or the true index of the relation which the faculties of the mind sustain to each other with respect to strength. The *form* of the head or brain does not give the whole character of the mind; it only gives the relative power of the several mental faculties. The form of the head has nothing to do with the *absolute* power of mind; it only determines the *relative* power of the several faculties. *Absolute* power depends upon other things. Two individuals may possess heads of exactly the same *form*, and one may be a simpleton and the other a genius of the rarest power and grandeur, and yet not in the least disturb the truth, that the *form* of the brain is the evidence and measure of relative power among the faculties. The relation of the several faculties in the simpleton's mind will be the same as the relation of the several faculties in the mind of the genius. Power depends upon other things.

Form affects the *balance* of power, but does not give *absolute* power.

The *form* of the brain is determined by the relative size of the several organs. The larger organ will always exhibit the stronger mental faculty. The comparative sizes of the several cerebral organs will determine the comparative strength of their respective mental faculties. Out of this grows the doctrine that "size of an organ is the measure of power, other things being equal." In the same brain the size of an organ is always the accurate standard of relative power, for the same conditions attend all the organs. But in different heads size is not a true measure of power, for the other material circumstances affecting power are not always, or scarcely ever, the same. These circumstances must always be attended to in estimating absolute power.

The doctrine that "size is the measure of power, other things being equal," is most amply attested. Take any number of human heads, attended by the same general conditions of health, temperament, and cultivation, and the larger ones will always be the stronger, and the degree of difference in size will tell the degree of difference in strength.

Look at the great men of all ages and nations—the men who have moved the world as though an earthquake's power resided in their wills—and they will be seen to have large heads and massive brains.

Take any one head, where you find one lobe or part of the brain much larger than any other, and you will find a corresponding strength in the mental faculty it exhibits. Take all animals, and it will be found that those which have the greatest amount of brain in proportion to their size, will manifest the greatest degree of mental acumen. The fox, the Newfoundland dog, the beaver, the monkey, and the elephant are among the best examples. Take the various races of dogs, examine the olfactory nerves of each one, and those that have the keenest, strongest scenting power, will be found to possess olfactory nerves as much larger as their power is stronger than that of other dogs. The olfactory nerve of the bloodhound is remarkably large.

The optic nerve in the eye of the eagle exhibits, in its great size, that bird's extraordinary power of sight. The same doctrine is found true in the cerebral, or nervous system, that obtains everywhere else in the animal economy, that the larger the organ the greater its power. It is so with the bones, muscles, glands, and every other part of the body. Why should it not be so with the brain? It most evidently is. But the practical phrenologist, especially the tyro in the science, must use great caution in his examinations and conclusions, or he will get greatly deceived by this doctrine. Power depends not upon size alone, but upon many other things in connection with it. He who judges of the physical strength of men by their size alone will often get greatly deceived. For it is not unfrequently found that smaller men are stronger, and they are generally capable of performing more labour, from the fact that they are more finely, closely, firmly organised. The texture of their muscles and bones is much more refined, compact, and perfect. Their muscles are often, in comparison with those of the larger men, like threads of silk in comparison with strings of tow. So he who judges of the strength of an *animal* simply by its size, often gets deceived. It is not always the largest horse that has the most power in him. Very much depends upon his make, his physical perfection, the closeness, compactness, and refinement of the texture, of his physical organs. But nevertheless, the general doctrine is true, that the larger the man or animal is, the greater is his strength—a fact, an important fact, it will be well to mention here.

As a general rule, it is not the strongest minds that will accomplish the most in the world. It is not the largest brain that will perform the most labour. It is not the largest man that will do the most work, nor the largest horse that will perform the most service. It is a philosophical principle in mechanics, that what is gained in power is lost in velocity. So in metaphysics, it is generally true that what is gained in absolute strength of mind is lost in activity. Daniel Webster had a mind eminent for power. It was a full-grown giant of magnificent proportions ; but it seldom used its power, in all its majesty of strength, oftener than once a year. It was usually slow but grand in its movements. And it was only when the stimulus of a vast combination of the mightiest circumstances poured their flood of strife and fire around him, that his mind was fully awakened, and all his resistless energies were summoned to the field of labour. Then it was that he outstripped all competitors, and soared in lofty grandeur into the mid-heaven of intellectual pre-eminence, the peerless giant in the sublime arena of mental strife, the just pride of America, and of the world.

John Quincy Adams had a mind not originally marked with

extraordinary power, but with excellent balance, and great activity. He could work almost at the top of his strength day after day and year after year, and accomplished more every year of his life than Webster ever did. He is nearer the mcdel man, an object of far greater admiration, a more beautiful and truly grand exhibition of humanity in its exaltation, than Webster could have been, had he lived ten lives such as his was. The name of Webster will be a tower of strength, but the name of Adams a dome of glory through all generations.

These two great characters illustrate the position, that it is not the mind of the most absolute strength that accomplishes the most in the world. *Activity of mind and endurance of mental effort* are as important as strength. These depend upon *conditions* of the brain, and not upon size. These *conditions* of the brain must be studied with greater care than any other subjects relating to the science, for they give tone, aspect, character, position in the scale of excellence to the whole mind. That peculiar possession of a singular and brilliant power, which the world knows under the name of genius, is more often given through some of these *conditions* than it is through size of brain. Geniuses oftener possess brains of only ordinary, and sometimes inferior size. But in these instances the brain is of the very highest order of texture. It is made with the most exquisite finish; refined and delicate as a model of perfection, quick as lightning, impressible and sensitive to the last degree.

The conditions may be known by outward signs, so that the *character* as well as the *size* of the brain may be determined with a great degree of accuracy by the close observer in phrenological science.

These conditions we shall endeavour to illustrate in the next lecture. At present there are other topics demanding our attention.

Does the outward form of the head show the true form of the brain? The negative of this question has been stoutly maintained by some claiming to understand physiological science. Why should not the head show the form of the brain? The brain takes its conformation originally from the character of the mind it is to serve. It is mind that gives it its form. It is mind that moulds it. And the mind existed before the brain existed—existed at least in the parents. The brain serves the mind, so it takes on the form that mind gives it. Now, the skull or cranium serves the brain. Its office is that of protector. It has no other use. It conforms exactly to the brain. It grows around the brain after the brain has taken on its full and perfect form. It is formed by a deposition of particles on the outside of the brain, and is at first a soft, yielding substance, lying closely around the brain. It forms around the brain something as the shell forms around the snail. It begins to form at several places at the same time, at about the centre of each of the bones of which it is composed, and extends in every direction till they meet and clasp in their embraces the whole brain, joining hands, and forming at their meeting-places the several sutures. Now, why should not the cranium show the form of the brain? Does not the skin which grows around the whole body show the form of the limbs? Most surely. But the skin is not of the same thickness in all places. In the most exposed places it thickens up to protect all the better those places. But we know just where those places are, and are not deceived by them about the general form of the body. So the cranium in the most exposed parts of the head grows thicker to afford a better protection. But these places we know, and are not deceived about the general form of the brain. Across the forehead and the back, and along the sides it is a little, and but a very little, thicker than on the top, and lower down. As a general thing it varies but little in the same head, and is from an eighth to a quarter of an inch in thickness. Where it covers the largest and most active organs, it is always thinner than where it overlays smaller and less active organs. The continued activity of the stronger organs causes them to grow and press out against the cranium, and this occasions an absorption, or displacing of the particles, which causes the cranium to diminish in thickness; while over the smaller and less active organs the cranium thickens by more full secretions, occasioned by the inactivity of the organs. The experienced

phrenologists will generally find but little difficulty in deter-
mining which are the active organs. The cranium will rise or
swell over them, and the swell will be more or less intense, or
abrupt, as the organ is more or less active. In cases where
the organ is large and has been subject to great intensity of
action, the outward prominence is distinct and sharp. In these
cases the cranium is very thin, often not thicker than a pen-
knife.

Take an empty skull and hold a candle in it, and it will
actually shine through those parts which overlaid the most
active organs of the brain that once occupied it. I once saw the
skull of a most abandoned and wretched woman, who had three
passions, to the gratification of which she gave her whole life.
They were *lust, anger,* and *music.* The skull over the organs of
Amativeness, Combativeness, and Tune was scarcely thicker
than a wafer; so that really the brain varies in its form a
little more than the outward skull, but the active organs and
this variation can always be very correctly determined. So can
the thickness of the skull be very generally determined. It is
thicker in people of a coarse, rough, bony make, and thinner in
those of a more refined, delicate, nervous constitution. Place
your hand on the head of an individual, pressing with con-
siderable force, and then ask him to speak. If his skull is very
thin his voice will jar his head in a perceptible manner. If his
skull is thick it will jar it much less. In this experiment the
character of the voice must be noticed. If it be heavy, a
greater vibration would be made on a skull of given thickness
than a light voice would make.

There are several protuberances on the skull which must not
be mistaken by the novice for organs. There is one on the
occipital bone, which is merely a bony process for the attachment of a muscle. It is called by physiologists the "spinous
process ;" by phrenologists, the "occipital spine." It is situated
just above and behind the upper vertebræ of the neck, above
Amativeness and below Philoprogenitiveness. There is another,
called the "mastoid process," situated just behind the ear. The
two may mistake it for Combativeness, though these processes

are entirely different from the appearance of organs. The processes are sharp and angular, the organs are gentle swells, or obtusely rounded elevations. We would caution every student of Phrenology against looking for *bumps* or protuberances, as in a well-balanced head these are not found. We calculate the size of organs by the distance from the centre of the brain, or from the head of the spinal column to the surface of the head at the location of the organ to be estimated. All the organs may be large, and the head without any special prominences. If one or more organs be large and others small, then we find hills and hollows.

The cranium is composed of two plates, the inner and the outer, separated by a spongy, porous, bony structure. At the sutures, or meeting places of the different bones of the skull, these plates are often more distantly separated. But this is usually distinguished by a sharp, angular elevation, extending along the line of the sutures. There is still another place where the outward form of the skull does not indicate the form of the brain. This is just above the roots of the nose, at the lower part of the forehead. The inner and outer plates of the skull are separated, leaving a space between, which is called the "frontal sinus." It sometimes extends sidewise under the arch of the eyebrows. It is not always easy to determine the size of the *frontal sinus,* but it is generally larger in persons of a coarse, bony make, and smaller in those of a more compact, refined organism. This sinus is generally small in the female head, nor does it ever appear in either sex until about the twelfth year, so that it offers no impediment to the estimation of the organs of children.

This includes all the bony protuberances which cause the outside of the cranium to vary in form from the shape of the brain. There are some parts of the skull covered so deeply with the integuments and muscles as to make it somewhat difficult to determine the shape of the head. About the temples there are thick and strong integuments, which serve to attach the lower jaw, which hide the true form of the head, though with a little careful observation and experience the general contour of this part of the head may be learnt with much accuracy. With these exceptions, the outward form of the head is an index to the form of the brain, so that in reality the head is the index of the mind. Every man has a chart of his soul on his cranium. His mind is mapped on the outer surface, for the world to behold and read; not really his mind, but a picture of his mind. His real, living character is written there in the autographic lines of God's own hand, so distinctly, indeed, that he who runs may read. His intellectual power and peculiarities, his moral tastes and characteristics, his social feelings, are all accurately described in the hieroglyphic characters of bone and brain. We have but to read this living history of the man, to know who and what he is. Our fellows, then, are not concealed from our view. They are not shut up in prison, where we can never know anything of them. Neither can they shut themselves up. The outward man always speaks of the inward. The physical man is moulded and controlled by the spiritual. The physical is the servant of the spiritual. Hence not only the form, and shape, and texture, but the motions, gestures, looks, tones, step, bearing of the outward person speak of the man within. There is not an action or aspect of the external man, not a smile or a frown, not a sigh or a laugh, not a light or a shade, not a song or an oath, not an expression of the face nor an action of the limb, that is not the result of a mental action or state. The mind is the king, and the body is its prime minister. Its first servant is the nervous system. The rest of the body is the servant of this system. So that the whole body must be learnt in order to learn the true index of the mind. The whole body is the index. The head and face are the most important parts in this mental research, but they are by no means all. Phrenology is really the study of the outward symbols of mind, the study of the mind's *language*, not the mind itself. From its language it is true we cannot well help drawing conclusions concerning the mind itself. But the proper study of Phrenology is the study of the menta. language written in and on the outward man.

As the body is the servant of the mind, it becomes necessary that it be sound, well formed, healthy, pure in its life and actions, else its service will be marred, distracted, uncertain, and impure. Little dependance can be put upon a weakly and corrupted servant. His whole surface will be tinctured with the jaundice or fever of his disease. So if the body is diseased, it will not, cannot serve the mind well. There is no moral lesson that Phrenology urges with more force and earnestness than that health—perfection of body—is of the utmost importance to our mental well-being. It has no fellowship with that doctrine which would crucify the flesh, abuse and corrupt the physical house in which we dwell. That house is the palace of earth's noble lord, and should be garlanded with the roses of health, and robed in the blushing colours of beauty. It should be an object of our tenderest care and solicitude. We should no more transgress a law of health than we should cut the throat of our neighbour. As we value mind, as we prize moral magnanimity of soul, as we estimate the glorious affections which bind us in links of gold to God and man, so should we regard the health and perfection of the body. Soul and body are joined in holy wedlock. They are a united pair. If one suffers, the other must. If the body decays, the mind cannot exert its powers. If the body sickens, the mind cannot use its appropriate powers, its appropriate language. Every faculty of mind has its outward, visible language. On the skull is written the strength and power of each organ, and consequently each faculty, and on the countenance and in the actions is written and spoken its natural, every-day language. Each organ has its own peculiar and appropriate language, different from all the rest. The organs of the mind's actions, may be compared to the great confederacy of nations. Each nation has a language, manners, customs, modes of action and expression peculiar to itself. So it is with each organ. The study of these several and varied languages constitutes one of the most pleasing and instructive departments of phrenological science. It is in these graceful and natural languages that human nature is daily exhibited, that the mind's peculiar phases, attitudes, and states are shown, that all the strange freaks of feeling and fancy are pourtrayed, that passion writes its burning words, that lust uses its bandy tongue, that anger thunders its annihilating threats, that love whispers its silvery notes.

No mental exercise is more truly delightful than reading the natural language of mind as it is written in the lives and actions of those around us. It is a knowledge of this language that enables us to read character, to study both ourselves and our fellows, to go in, as it were, into the sanctuary of their souls, and sit in meditation there when they know not what we are doing, to examine the actions and states of their minds, and make ourselves acquainted with them as they really are. It is in this language that is written the highest and grandest actions of mind, such as the philology of the tongue and pen can never express.

We often have ardent aspirations, burning loves, overpowering sorrows, uncontrollable joys, intense devotions, lofty thoughts, to which no human language can give adequate expression, so that the best, the loftiest, the grandest views of the human soul can never be painted on canvas, or spoken in words. It is left for the natural language of the organs of which I am speaking, to utter in our presence, and pourtray to our eyes, those splendid flights and burning feelings of the mental man. It is the language, and the only language, in which the real, living poetry of the soul is written.

Byron has told us well how we are often left to the use of this natural language to express our thoughts and feelings. Says he—

> "Could I embody and unbosom now
> That which is most within me; could I wreak
> My thoughts upon expression, and thus throw
> Soul, heart, mind, passion, feeling, strong or weak,
> All I hear, know, feel, and yet breathe, into *one* word
> And that word were lightning, I would speak."

But as it was, he found himself unable to utter the burning lava-tide of feeling to which his soul had risen. Could he have been *seen* then, the *natural* language would have spoken the sublime poetry of his mind, and poured out in one rich, full, flaming expression, the lightning thoughts that were wrapping in a blaze of glory the canopy of his soul. It is the free use of this natural language that gives the actor and the orator their power, that is the soul of eloquence, the poetry of life, the spirit of all mutual influence and power.

This language it is the province of Phrenology to teach, so far as it can be taught. Yet only its plainest and commonest forms are all that can be *taught*. It must be learned by observation, by the most critical attention to the natural modes of expressing feeling and thought. As we pass along we shall speak of the natural language of the several organs, as far as time will permit.

It is proper that I should call your attention for a few moments to the brain. I propose not to detain you with a long dissertation upon the physiology of this important centre of nervous power. A general outline is all I can think of giving. The brain is composed of a soft, yielding substance, nearly destitute of anything like fibres or texture. It is thoroughly supplied with blood-vessels and uses about one fifth of the blood of the system. It is divided up and down into two lobes, or hemispheres; so that all the organs are formed in pairs, as the two ears, eyes, hands, feet, &c. It is divided horizontally into what is called the cerebrum and cerebellum. The cerebrum is the main body of the brain, and is above the cerebellum. The cerebellum is the base and back part of the brain, lying close down upon the neck. It is separated from the main brain by a thick, strong membrane. In some animals, particularly those that leap for their prey, it is separated by a thin partition of bone. The cerebellum is composed of material very, if not exactly, similar to that of the cerebrum, with which it unites at the common centre, just above the top of the spinal column. The various nerves, both of motion and sensation, from the whole body, meet in this place. Here is the common centre of all the organs of the brain, and of all the nerves of the body. To this centre go directly the optic, olfactory, auditory, and gustatory nerves; and here, exactly at this centre, are the two halves of the brain and the nervous system united by a strong band of nerves, or a large single nerve, which forms, as it were, the hymenial band between the two otherwise distinct persons. It makes them literally one. Were it not for this there would be two sets of feeling, sensation, motion, ideas, emotions. If we looked at an object, we should always see two. If we heard a sound, it would be double. But this great matrimonial nerve unites the two halves of the nervous system, and blends all their thoughts, feelings, emotions, and perceptions into one, so they think, and feel, and act as one person. The double ideas and perceptions that come in from the outward world are all formed into single ones by this nerve, placed here in the grand centre of all nervous power. It has been suggested by someone, that this is the proper dwelling-place of mind, its *sanctum sanctorum* its perpetual presence-room, where it lives, acts, feels, and from which it issues its mandates, and sends out its thousand living voices to sound around the world. Of this we cannot know, in the present state of knowledge; but the theory is a beautiful one, and highly charged with probability.

The surface of the brain is marked with convolutions, resembling very irregular folds, which serve for its enlargement. By this means the outward surface is much more extended than it otherwise would be, for every time the surface is folded in it doubles just to the depth of the fold the extent of surface. It is more than probable that the power of the brain depends upon the extent of its surface. We know that galvanic power is always in proportion to the extent of the plates employed, and if the brain is a galvanic battery, as was suggested in the first lecture, then the idea is not improbable that its magnetic power depends upon the extent of its surface.

These convolutions vary in depth in different brains. In some they are very shallow, in others they sink down deep into the substance of the brain, so that ofttimes the smaller of two brains possesses the greater amount of surface. Then if strength of brain depends upon extent of surface, the smaller brain would exhibit the more mind. We know this is often the case; and the reason for this may perhaps be explained in this way. I am of opinion that these convolutions are of immense importance in the cerebral economy, that their depth determines the depth or strength of mind, so that if we could actually measure the surface of the brain we could measure the amount of mental power.

The question then arises, Are there any outward signs or indications by which the depths of these convolutions may be determined? This is a subject yet open for investigation. But in the present state of enlightenment it is rendered more than probable that the temperament will give us approximate, if not very accurate, knowledge upon this subject. The more refined, delicate, compact, and nervous the physical constitution, the deeper the convolutions, the greater the extent of surface, and consequently the greater the mentality. It is probable that precocious children and youth, geniuses of rare and general powers, *prodigies* in intellect, have great depth of convolution in their brains. To this point the inquiry of all phrenologists should be directed. Brains of known and remarkable power in life should be examined critically in death. A long and faithful comparison of known mental power should be instituted with the brains which exhibited it. The truth in this matter may eventually be reached, and when it is, there is little doubt but that phrenological science will be a mathematical rule.

The lower order of animals have no convolutions of brain; but as we follow up the scale of animal intelligence they appear, at first indistinctly, and become deeper and more numerous as we rise to the dog, the horse, the elephant, and the ape family. The brains of the most intelligent of men, like Cuvier and Byron, have been found, on dissection, to contain convolutions double the depth of those of moderate mental capacity.

To another fact most grand in its practical bearing, I will call your attention. It is this: the *exercise* of each or any organ causes it to expand and become both more strong and active. Any portion of the brain that is rigidly and strongly put to labour will acquire an increase in size and strength by that labour. The general law holds good here which is applicable to the muscles, the nerves, the glands, or any other

portion of the body. The blacksmith's arm acquires its huge
dimensions and giant strength by the repeated strokes which,
day after day, and year after year, it is called to give. The farmer's hand is made large and powerful from a similar cause.
The nerve of one eye is increased in size and strength when
the other is destroyed, in consequence of its increase of
labour. The auditory nerves and the nerves of touch become
large and intensely active when the sight is lost, because they
are called upon to perform the labour of another sense. This
general law applies with all its force and beauty to the brain.
And it is by the force and utility of this law that the science
can be made most eminently practicable in balancing, harmo-
nizing, and perfecting our mental natures. If any portion of
the brain is too small, it can be whipped into the traces, and
put vigorously at work till it acquires both the strength and
activity of the other portions.

If any number of organs are too weak, they can be thus
strengthened. By a critical self-examination, which every one
should daily make, we can discover our weaker organs, and
apply the only remedy. We can discover the notes of inhar-
mony in the mental anthem which we are every day chanting,
and key the instrument of our souls into tune. When harmony is attained, when a balance of mind is secured, when all
the organs are of equal strength and activity, then with us
the millennium has come, the day when the gates of joy and usefulness will be thrown wide open, for us to enter the kingdom
of righteousness and peace.

LECTURE III.

Temperament as affecting the Quality and Power of Mind—The Physical
the Measure of the Mental—Mind gives mould to Matter—Difference be-
tween Man and Woman—Man stronger, Woman more intense—Refine-
ment, a source of Mental Power—Woman subject to Extremes—
Channing, Josephine, Napoleon, Adams—J. C. Neal, Refinement and
Power combined—Poets and Thinkers contrasted—Effects of Equal
Power and Activity—The real Men of Action—Balance, the Perfection of
Humanity—True Philosophy of Marriage—Temperament Illustrated—
Bilious and Lymphatic Temperaments—Dulness of the Lymphatic—
Fire of the Sanguine—Mentality of the Nervous—Every man has a mixed
Temperament—New Theory of Temperaments—The Body the Casket
and Mirror of the Spirit.

IN the last lecture we spoke of size of brain as affecting absolute power of mind. We now come to the other conditions
then referred to, which affect absolute power. These conditions
are called " *temperaments.*"

The most casual observer of humanity has not failed to
discover that men differ vastly in sensibility, refinement, exquisiteness of feeling, intensity of mental action, quickness of
thought, vividness of perception, and in delicacy of sentiment
and emotion. Some persons are coarse, and rough, and uncouth
in all their mental characteristics. Their thoughts are rough-
hewn, ragged, jagged, uncomely, resembling boulders of granite,
fragments of rock, and are always expressed in language as
coarse and unpolished as themselves. Their affections are of
the same nature—rocky, harsh, outlandish, and their expression
of them equally so. Their moral sense bears the same marks
of rude, ill-defined, and coarse ideas of duty, devotion, and
holiness. Everything they think and feel; everything they
do, and say, and love, bears the mark of this peculiar rough-
ness. They always make us think of old chaos ; of the earth
before it had tumbled into form ; of a continent of mountains ;
of an ocean of billows; of a city of log-houses; of a new
settlement, where dry trees, and green stumps, and piles of logs

D

cover half the land; of a rudimental, or barbarous state of society, where everything is blunt, and coarse, and rough-hewn. Others there are whose thoughts and feelings are elevated, pure, refined as a note of exquisite music, delicate as the strings of an æolian harp, intense and fine to the last degree. Their affections have the same marks of an exquisite refinement, and a strong and lofty intensity. Their ideas of beauty, their emotions of sympathy, their sense of duty, their perceptions of harmony, their joys and sufferings, are all characterised by intensity, delicacy, and refined sensibility. They remind us of wisdom's embodiment; of love's ideal; of a perfected soul; of society harmonised; of the Spiritual kingdom established; of the resurrection state.

Between these two extremes there is every possible shade of character.

Now, is there anything in the physical man that will give us a correct idea of the mental character in respect to its intensity and refinement, or that will measure the degree of pure mentality? In the two extreme cases which I have given, there is a vast difference not only in the kind, but in the degree of purely mental power. The one is as much superior in *degree* as it is in *kind* to the other. Is this difference written in, or on, the outward man, so that we can read it? Phrenology says it is. And if the doctrine upon which phrenological science rests be true, then its teachers have good ground for their assertion.

That doctrine is, that mind or spirit rules and moulds matter. If so, then the constitution of the body will tell the constitution of the mind. The refinement and delicacy of the body will be the index of the refinement and delicacy of the mind; for the simple reason that the body is what it is, by virtue of the mind which moulded and dwells in it. The body is coarse because the mind which made it so is coarse, and has always used it for rude, coarse purposes. Or, the body is refined because the mind which made it so is refined, and has always used it for refined and delicate purposes. The body being subject to the mind, it must possess its peculiar character as the body, as a gift from the mind, as an inheritance bearing the peculiar mark of its original proprietor. Take the coarsest, roughest *man* in your knowledge, and the most refined and exquisitely wrought *woman* in your circle of acquaintance, and compare the two with respect to physical delicacy and refinement. Look at their hair. One is coarse and bristly ; the other is soft and fine as threads of gossamer. One is black as the hues of night, the other is golden as the radiant sunset. Observe their skin. The fibres or texture of one is as coarse and harsh as a web of crash ; those of the other as fine, smooth, and almost invisible, as the threading of a piece of the glossiest silk. Witness their hands, feet, and limbs. Compare them, not in size simply, but in the delicacy of their make, their form, their elegance, their fineness. How marked, how great the contrast! In every respect it is as visible and distinct as the variety of forms in the outward world. Now the difference in the outward persons, with respect to refinement and delicacy of constitution, is no greater, but just as great as the difference in their minds in this respect. The refined constitution will exhibit not only a more refined kind of mentality, but a greater amount, a greater intensity, a greater force of mind in proportion to the size of the brain. There is no doubt that the convolutions of such a brain are far deeper, and perhaps more numerous, and the intensity of its actions far greater and more powerful.

Again, observe the difference between man and woman— between women in general and men in general. Woman is far more delicately wrought and exquisitely formed than man ; and she exhibits a degree of mental power in proportion to the size of her brain, as much greater than man as she is more refined than he. Hence it becomes necessary that man should be larger than woman that he should have the same amount of power. There is no doubt that power of mind is about equally balanced between man and woman. What he lacks in delicacy and refinement of brain, he makes up in size. And what she lacks in size of brain, she makes up in intensity of tempera- ment; so the difference between them is not in power but in kind of mentality. Her system being more compact and refined than his, she is capable of more intensity of action, and possesses greater powers of endurance, in proportion to her strength. Hence he needed greater strength in order that he might be capable of doing and enduring as much as woman. The more compact, refined, and well-formed a human system is, the more it can do and endure, the longer it will live, the more it will accomplish, and the more healthy will be the products of its mental activities. This explains the reason why frail, delicate woman will often perform such wonderful labours, live under such enormous burdens, and endure such intensity and length of mental and physical sufferings. And this, too, shows the effect of physical refinement and perfection in affecting mental power. The physical difference between man and woman illustrates, too, this same principle. If man is larger, woman is finer. If man is stronger, woman is more intense. So that the great doctrine of the power and influence of temperament may be learnt by a contrast of man with woman, physically and mentally.

. The question has long been agitated, respecting the mental difference between man and woman. It has been contended that she is the weaker in intellect, because she is smaller and weaker in physical strength. But this argument will not be admitted by Phrenology; for that shows that real power depends not altogether upon size, but upon other conditions. If these other conditions which confer mental power are found in women, then the argument against her is not good. The whole female conformation shows that these conditions are amply made up in her constitution; so that her mental power stands side by side with man's. But here a question may arise,—Is the power conferred by *refinement* of constitution, which is woman's great source of power, the same in kind with that conferred by *size*, which is man's peculiar source of power? Is there any difference between the two? It is my opinion that there is. The power conferred by refinement of constitution is altogether a higher order of power. It is nearer purely spiritual power. It is by this that the highest order of intellects are formed. It is this that makes poets, artists, geniuses. It is this power that lights the flames of the purest and most intense intellectuality. It is this that gives that kind of intuitive intellect which sees with a spiritual eye, which comprehends without apparent reasoning, which darts through a whole subject with lightning rapidity, and which, seer-like, beholds the shadows of coming events cast before. It is minds formed by this power that have delighted and charmed the world. They have written its deepest, loftiest poetry; they have made its sweetest, intensest music; they have poured forth its most refined, touching eloquence; they have painted its liveliest colours and have chiselled its most perfect forms; they have breathed its holiest prayers; they have cherished its loftiest virtues; they have lived the most intense and glorious lives. Such minds dwell close upon the borders of spirituality. The life they live is half divine. They are human angels. A glory from above encompasses them. Their thoughts are electric spirit-flashes. Their loves are flowers of ethereal passion. Their devotions are reverent poems of praise and love of the Divine Spirit. Their emotions are music-strains of the most refined joy and grief.

Of this kind of power woman shares more largely than man. Hence, hers is a more intense and glorious life than his. Hers is a more refined and elevated character. She is better and wickeder than man. She is nobler and meaner than he. She is higher and lower; purer and baser; sweeter and bitterer; gentler and fiercer; lovelier and more hateful than he. That very power which will make her almost an angel when properly used, will make her almost a devil when abused. But that power she more frequently uses for good than evil. Enlightened woman turns it almost wholly to the heavenly side of her character, and hence is elevated close upon the precincts of angelic life. The degree of this power in woman over man is shown in the superior elegance, refinement, symmetry, and beauty of her physical system. Then the sum is this. Man has more of one kind of power; woman has more of another. Both kinds are equally useful and necessary in the life we now live. He who has too much of one kind of power is too much

of an animal to elevate either his own or his fellow's character.
He who has too much of the other is too much of an angel to
understand and know how to relieve the most pressing wants
of the mass of humanity. These two kinds of powers are,
strictly and philosophically speaking, the masculine and feminine powers of mentality. Man has more of the masculine;
woman has more of the feminine. That character is most
perfect in which these two powers are equally balanced, whether
it be in man or woman. It may not be improper to mention
some notable characters in which these powers seem to be well
united. Dr. Channing presents himself to my mind as having
had the most perfect balance of any man of general eminence
in my knowledge. I conceive that the two powers of which I
have been speaking, were very nearly equally balanced in his
character. Hence he exhibited not more of the man than the
woman in the character of his mind and the nature of his
feelings. He was powerful and tender, lofty and pathetic,
severe and sweet, grand and intense. He was at once the
noblest and purest, the sublimest and loveliest, the greatest
and best, of American men. His name and his character are
a living glory to the world. Ages hence, he will stand in the
firmament of enduring excellence, an object of wonder and
beauty for the admiration of the great and good.

The Empress Josephine, the first wife of Napoleon, is the
best example among women to which my mind now reverts.
She was by nature a great and good woman, a paragon of spiritual excellence and beauty. She possessed about an equal
share of each kind of power, and hence lived both a powerful
and intense life. Napoleon possessed more of the power given
by *size*, more of the masculine, though the balance was not
greatly in favour of this. He had a strong development of
both kinds, hence he was one of the most intense as well as
most powerful of men. John Q. Adams had a little more of the
masculine, or the power given by size, though in him the balance
was but little disturbed. Joseph C. Neal, the author of the
' Charcoal Sketches," had as perfect a balance of power as we
often find. In the characteristics of his spirit he was about as

much of a woman as man. He had the intensity of woman and the power of man united. Perhaps there never was a man who, simply by the power of the pen, attained such a wide-spread popularity, both among men and women, in so short a time, as did he.

The poets Cowper and Whittier are of this equal combination. Mrs. Hemans and Mrs. Mayo had a little more of the feminine than the masculine kind of power. Webster, Corwin, Benton, Cass, had a strong predominance of the kind of power given by size. If you will examine the physical structure of persons in whom these two kinds of power are equally balanced, you will find that they are refined, compact, firm, and capable of great endurance. They can endure more intense labour, both physical and mental, more suffering, more excitement, more exertion of body and mind, than any others. They are both strong and active, quick and powerful, in body and mind. They are wiry, tough, hardy, and supple. They are not so powerful in physical strength as some others, but what they lack in strength, they more than make up in activity and power of endurance. They seem to work easily, with little fatigue or effort, either bodily or mentally. They go like a perfect machine, without fatigue or friction, jar or discord. Hence there is no waste of strength, of energy, or time. They make the most of everything, live the easiest and fullest lives, perfect most their natures, accomplish most in the time allotted them for this sphere of existence, and generally live to the greatest age. They are generally moderate in size, passing to neither extreme of high or low, of large or small; of moderately fair complexion, neither very florid nor very pale, very light nor very dark; of limbs, muscles, and form, full, well rounded, yet not extremely so; full chests; erect in stature; of heads proportioned to the size of their bodies, and general symmetry of person. They are generally healthy, and equally capable of mental or physical labour. They can resist extremes of heat and cold, live in any climate, perform any kind of labour, live under any bearable circumstances, make everything tell in their behalf that can possibly be wrenched into their service, can dig success out of rocks, misfortunes, opposition, trouble, and make almost everything count in their favour. They are generally up with the times, ready for bargains, opportunities, chances, openings, or whatever will be available. They can suffer and work on, rejoice and forget not the object of their pursuits, be excited and not thrown off their balance, be frightened or shocked and not lose their presence of mind, be greatly tempted and still resist, be opposed and not overcome, beaten and not conquered, coaxed and not seduced. In a word, the balance-wheel of their minds and bodies seems never to vary very much from its regular and proper motions. Such persons are the most reliable, safe, useful, sure, of any that can be found. And these general qualities, running as they do through all the various departments of human thought, feeling, and action, are given chiefly by the temperament, by a proper or equal union of the two kinds of power conferred by size and texture, or the two which I denominate the masculine and feminine powers of humanity. I give them these names because one is generally found predominant in man and the other in woman. And it is the predominance of each that gives to both man and woman the peculiar characteristics for which they are each remarkable. Man is superior in the power given by size ; hence he is man, or possesses the nature that we ascribe to the masculine. He is larger and stronger in a certain kind of strength. Woman possesses more of the power given by texture; hence she is woman, or possesses the nature that we ascribe, or have found to belong, to the feminine. But it must be remembered that every man and every woman possesses both these powers to a greater or less extent. We are to study the masculine nature, or the evidences by which its presence is tested in every individual, whether male or female; to study the feminine nature, or the signs of its presence, and then determine the relations they bear to each other, before we can determine the peculiar nature of any person's mind. It is the perfect balance of these two that constitutes the perfection of temperament ; and it is the perfect balance of all the mental faculties, or cerebral organs, united with a balance of temperament, that constitutes the perfection of humanity.

It may be objected by some, that these views are too theoretical to be of practical value. But instead of being purely theoretical, they are founded upon the peculiarities which are known to exist and be visible in the two phases of humanity, as exhibited in man and woman. They are founded upon the universally admitted physical and mental natures of man and woman; recognise and account for the acknowledged difference between the two sexes; explain many of the peculiar likes and dislikes, or attractions and repulsions, which are everywhere exhibited so strongly as long since to have passed into common sayings. There is a natural tendency in all things to an equilibrium. This law holds as good in mental as physical philosophy. Toward this equal balance of which I have been speaking, all minds are tending. The attraction which operates most strongly upon them is always toward perfection, or from a direction opposite to their imperfections. This is strikingly exhibited in the likes and dislikes of men and women for each other. A very tall man likes a short woman, a very tall woman prefers an opposite kind of man. A very corpulent man or woman admires for a companion an opposite physical conformation. The same general law is true in all cases of extreme temperament. There is always an attraction from an opposite direction from the direction in which perfection is to be found. Perfection lies in a balance of the two powers which give a perfectly moulded form. It seems that mind has an instinctive idea of this fact, and hence is always attracted toward this centre. Now, it is very easy, or, to say the least, not an insurmountable task, to determine, from what we can learn of the peculiarities of the two sexes, what are the masculine and what the feminine peculiarities, and what constitutes a union of the two. If we can readily learn these, then the theory is perfectly practicable— we can apply it in every instance; and it will be found upon a little careful examination, that in no other way can temperament be examined so successfully, and applied to actual observation so accurately. We can learn so readily what is masculine and what is feminine, and what is the medium between the two, that we can judge with great accuracy the peculiar mental characteristics that are conferred by any human temperament.

If I have said enough to give you my idea of the two kinds of power, and the way in which they affect temperament, I will call your attention to the divisions of temperament made by the most eminent phrenologists. They are *four;* and are named from the four great systems in the corporeal economy ; viz., the osseous, or bony; the circulatory, or sanguineous ; the digestive, or nutritive; and the nervous systems. In all human forms these four systems are combined, sometimes in equal or perfect proportions, and sometimes in very unequal or imperfect proportions. They each perform a particular office, and exert a peculiar influence. The first is called the *" bilious temperament,"* and is named from the osseous system. This is the skeleton or framework of the body. Much of the strength and durability of the body depends upon the excellence of this system. It is this which sustains the weight of the body and bears its numerous burdens. When this temperament is properly developed, it gives a full, fair-sized, well-formed, and well-proportioned frame. The bones are neither too large nor small, nor the joints too clumsy, nor the frame too heavy nor light. When it is strongly developed, so as to give its peculiar marks, it gives a dark, heavy, lowering aspect to the countenance, by its large, arched eyebrows; large nose ; high and prominent cheek-bone ; coarse, black hair ; large, black eyes; rough, bony forehead ; and heavy chin. The bones are large and angular; the joints large and rough; the whole framework strong and coarse. The complexion is dark, and the skin exhibits a somewhat coarse organisation. It gives slow, heavy, awkward motions to the body, and confers strength and powers of endurance. It is slow to move, slow to work, and slow to get tired. It is always best on a long race, and in the afternoon. It is the all-day temperament. It is powerful but slow. It gives to the mental actions the same peculiarities that it does to the bodily—coarseness, awkwardness, slowness, and power. It is often found in some of the greatest and most powerful of men, united with *good* sanguine and nervous temperaments. Daniel Webster and Thomas Corwin are perhaps two of its best examples. Men of this temperament are seldom found in the higher ranks of literature, art, or science. They are formed for power, but not for those nice, fine, keen perceptions which are necessary for the highest walks of life. If they are men of power, they are generally found in the field of political or military strife. Men of this temperament can bear burdens, losses, misfortunes, opposition, well; because they do not feel so acutely and sensitively as those of a different organization. Still when anything does affect them, it affects them strongly, and they have not that elasticity of spirit which others often have, to throw off a load of oppression or despondency. They fail in buoyancy and elasticity of mind. They are permanent, firm, and enduring in power and feeling.

The second is the " *lymphatic temperament*," named from the digestive or nutritive system. Every one knows that digesting is the enemy of thinking and feeling, that the mental processes are in a great measure paralysed by the digestive processes. Hence the lymphatic temperament cannot be considered a mental temperament; it is rather a physical one; and when it predominates we can seldom look for great mentality. Its outward signs are fullness and rotundity of form and limbs, wide, thick, leaden, inexpressive features; thick lips; round, blunt chin; light complexion, thin, soft, straight, rayless hair; light grey eyes; soft muscles; coarse, soft skin; with a relaxed, unstrung, loose appearance to the whole system. It is the office of this temperament to supply the waste occasioned by the mental. Hence, instead of working, it proposes resting; instead of thinking, it prefers sleeping; instead of excitement, it loves calmness. Instead of anything severe, intense, or active, it chooses a lazy, lubberly laugh. It is the slip-shod-and-go-easy temperament, the eating and sleeping temperament, the feeding and fattening temperament. It is dangerous to predict intensity, activity, mentality, spirituality, when we find this temperament strongly preponderant. It makes good-natured, easy, quiet harmless people. Yet there are sometimes strong minds connected with this temperament, but they never hurt themselves with work. They go to bed early, sleep soundly, and rise reluctantly to a late breakfast, which to such good feeders is the strongest temptation to seduce them from their slumbers. Their mental perceptions are generally dull and cloudy, and their actions all sluggish.

The third témpérament is the "*sanguine*," named from the blood. And as the blood is the furnace of the body, and carries the fire and flame by which the whole is warmed, it is but natural to suppose that this is the warming temperament. We read about "hot bloods." They are the people in whom this temperament predominates. It is the burning, flaming, flashing temperament. Hence it hangs out its signs of fire in its red, blazing hair and countenance, its florid or sandy skin. It has blue eyes; round, full features; pliable, yielding muscles; full, ample chest; generally, a thick, stout build; sometimes chestnut hair. It gives activity, quickness, suppleness, to all the motions of body and mind; great elasticity and bouyancy of spirit; readiness, and even fondness for change; suddenness and intensity to the feelings; impulsiveness and hastiness of character; great warmth of both anger and love. It works fast and tires soon; runs its short race and gives over. It is fond of change—light, easy, active labour; fond of avocations that require but little hard labour, and much of out-of-door jollity. It loves excitement, noise, bluster fun, frolic, high times, great days, mass meetings, camp meetings, big crowds, whether for religious, political, or social purposes. It is always predominant in those active, stirring, noisy characters that are found in every community. It loves with a wild intensity, but gets over it soon, when deprived of the stimulus afforded by the *presence* of its object. It feels grief and sorrow most bitterly, but soon becomes calm and forgets it all. It confers the most perfect elasticity to the mind, and the sprightliest bouyancy to the spirits. It makes warm friends and fiery enemies, and they may be both friends and enemies in the same day, and be perfectly sincere. It has a ready tongue; is quick and sharp of speech; is full of eloquent flights and passionate appeals; is ardent, pathetic, and tender, to the last degree : can cry and laugh, swear and play, in as short a time as it would take some people to think once.

The fourth temperament is the "*nervous*" and is just what its name indicates. It is given by the nervous system, and is emphatically the *mental temperament*. It is this, and this alone, that gives mind. The others affect the manifestations of mind only as they modify the actions of this. As the nervous system is connected with, and related to the other systems of the body in the most intimate manner, it must be affected more or less by them. But it should be remembered that they affect mind only as they modify the actions of this temperament. The nervous system is the mental medium. When this system is strongly predominant it gives the countenance a strong expression of intellectuality, a deep, clear, serene thoughtfulness, a brilliant dawning of mentality. It generally is shown in light, fragile, active forms; narrow, flat chests; tall stature; large head in proportion to the body, the upper part of the head being the larger; light complexions; thin, fine, glossy hair, usually quite light in colour; blue, or hazel eyes; thin lips; sharp nose; narrow chin, or a sharpening of the lower part of the face; a clear, transparent skin; small neck; small, yielding, flexible muscles; often a stooping posture; and a general lightness and gracefulness of motion. It gives clearness, precision, and activity, to all the mental perceptions; seeks mental pursuits, rather than physical; thinks, loves, aspires, with great ardency and devotion. Its joys, pleasures, griefs, sorrows, all its feelings are indescribably intense. It enters heart and soul into all it does; is permanent in its mental states, always the same, ardent, devoted intense intellectuality. It is the poetic temperament, and fills the mind with the flames of poetic fire. It sees and feels everthing under a poetic aspect and character. Its feelings are all ardent passions, and they burn within it like deep, subterraneous fires; yet they are generally of an elevated character. It is the temperament which makes angels on earth; which gives us an idea of angelic feelings, aspirations, and affections. The states of mentality to which it will elevate its possessor are altogether indescribable. It is the temperament which makes geniuses, precocious children, people of purely intellectual habits and tastes. In one word, it is the *mental temperament.*

It may be observed that these temperaments are always all found in every individual. No one can exist without them. They are the outward manifestations of the strength of the internal systems. Their combinations are as varied in different persons as their forms and features. It is not often that two can be found just alike. The character is greatly affected by the combination; so that the utmost care should be taken in obtaining a correct understanding of the temperaments. These temperaments have been called by some phrenologists the "Motive," "Vital," and "Mental," temperaments; the "motive" corresponding to the *bilious,* giving strength and energy of character—strong motive power; the "vital" corresponding to the *sanguine* and *lymphatic,* giving active life, energies, and the warmth and glow of a superabundance of the life principle; the "mental" corresponding to the *nervous,* giving pure mentality. This classification is much preferable to the other for practical purposes; yet I regard the other as much more purely scientific, and more readily comprehended by the tyro in the science. •

Yet I regard the view which I gave at first of the two kinds of power, one given by size and the other by refinement, or the masculine and feminine principles of humanity, as more practical than either of the others, and thoroughly scientific. It will lead us to new modes of investigation, open to us new views of our common nature, and explain many of the daily phenomena of mental character, which otherwise are but darkly understood. Still we should familiarise ourselves with all these views, for they are but different phases of the same general principles on which phrenological science in a great measure depends.

Nothing can be more delightful than the study of temperaments; for the student very soon accustoms himself to associate with any given temperament, the peculiar mental states which it confer or predisposes to; and thus he comes into almost immediate contact with mind. He reads mental language, mental characteristics, mental modes and forms of speech. He associates himself and all outward forms with mind; he looks upon the body and all its states and changes as mental effects— the result of mental states and changes; comes to regard the beings by which he is surrounded as spiritual, not as physical beings; sees, feels, converses, and associates with them as spiritual persons; loves, cherishes, and blesses them as such. He forms all his alliances, friendships, relations of mind; lives and dwells perpetually with mind, so that all his conceptions of men are elevated, spiritualised. Everything he sees in the physical man speaks of the spiritual man. Hence physical perfection, physical symmetry, beauty, gracefulness, carries his mind into the spirit out of which it grows; and he stands, as it were, in mute and rapt admiration of the spiritual being he beholds.

LECTURE IV.

Appreciation of the Works of the Creator—Beauty of the Science of Phrenology—The different Mental Groups—Position and Power of Organs—Offices of different Faculties—The Perfective and Moral Group—Man's Nature a Proof of God's Existence—The Domestic Faculties—The Selfish Faculties—Influence of one Faculty on another—Balance of Groups the Perfection of Character — Affectionate Group — The Desire of every Faculty a Love—Amativeness: its Office—Man alone, Imperfect—Purifying Effects of Amativeness—Curses of abused Amativeness—Proofs of degraded Amativeness—its Effects on Married and Single—Location of the Organ.

THE attention of the class is invited in this lecture to a careful consideration of some of the beautiful features of the grand and glorious science we are investigating. This science is like nature's scenery among the Alps, beautiful and grand at every view. As we pass along, the mind that appreciates God's perfect, sublimely perfect works, cannot but be filled with wonder and admiration. Yet notwithstanding all its intrinsic excellence and grandeur, the science itself teaches us that we cannot expect that all will or can at present appreciate it.

The subject of the last lecture taught us that some persons are organised for no higher aims than to supply and gratify the demands of the animal desires, to be satisfied with the pursuit of the coarse and the low, the vile and the vulgar. Hence, when they are called to admire the transcendent beauties of this science of all sciences, we can only expect that they will look on with a cold apathy, or turn away to talk of some vulgar sport, or to concoct some scene of animal lewdness or merriment.

Place some people amid the wild grandeur of the Alps, show them mountain peak rising above peak, as far as eye can stretch on every side, crowned in the flashing coronals of everlasting ice and snow, glittering in the cold sunlight like the heads of so many monarchs far up in the clear sky, while down their sides hang the solemn waste of impending glaciers a thousand fathoms above the vales, and summer is smiling below in rosy beauty at their feet, and they will look on with stupid unconcern, or turn away to gossip, or gormandize, as their vulgar tastes shall lead them. Some minds there are, however, to whom these Alpine views are a feast of glory, an intoxication of joy. Still nobler and higher are the minds required to perceive the excellence, and be electrified with the beauty of the mental scenery which our science reveals. I trust that I address myself to some such minds. It is the joy of my life, the glory of my being, to instruct and commune with them. To me they are earth-angels, prized, admired, and loved as such; and I behold a glory around them infinitely more splendid and dazzling than that which flashes in cold grandeur around the heads of the Alpine mountains. To me, they are living, progressive, spiritual immortalities, flashing from their brows the light of their divine Author and Guardian, in whose image they were created. The interest and affection which I feel in and for such souls approaches well-nigh to an extravagant idolatry. And those feelings are greatly heightened and strengthened by the illuminating and beauty-revealing power of this our truthful, God-written science. There is a divinity in this science, for God is its author and its primary teacher.

In the examination of the mental organism, the first peculiar

feature that strikes our attention is the association of organs
They seem to be grouped in families ; or each one seems to be
situated between the neighbours that are nearest its kindred in
their desires. It looks to me a little as though Fourier's principle of Association was pretty strictly observed in the arrangement of the particular organs, and in the arrangement of the
families or groups. It is quite certain that Fourier's primary
principle, and that on which the organs are arranged, are one
and the same. Whether Fourier got it from the natural constitution of man is a question we will not attempt to decide.

Here, too, in the arrangement of the organs, we find one of
the primary principles of Swedenborg, which is, that men associate and love on the principle of congeniality. And here, too,
in the great social structure of the mental family, we find the
primary principle of human perfectability, which is, that the
perfection of the great family is made up of the perfection of
all its members, or mathematically, that the whole is made up
of all its parts.

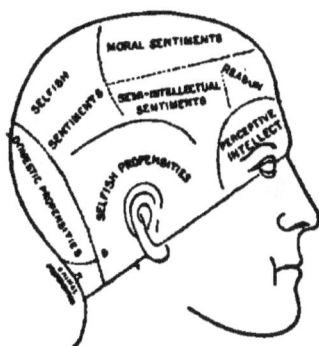

MENTAL GROUP.

(For arrangement of the organs, see the Frontispiece.)

Let us examine the mental grouping a little. Here, in the
frontal region of the head, as if to stamp on man's very visage
his intelligence, is the intellectual family, the ruler, father, or
president of which is Causality, or reason. In the centre

of this intellectual family dwells Causality.   Around it are
gathered its dependents, or the members of its family.   Nearest
to it, and just below, is Eventuality, the office of which is to
keep the treasure-house of the mind, or the treasures which
Causality or reason wishes to use.   Reason could not work,
would be useless, an imprisoned power, were it not for this
treasure-house, from which to draw the means it must use in
obtaining its results, its premises for every argument and conclusion.   Then, again, the treasure-house of Eventuality would
be useless were it not for the labourers which are necessary to
fill its store-rooms.   Its next-door neighbour below is Individuality, that industrious gatherer of all items, that universal
observer, who goes about with spy-glass and microscope, peep-
ing into everything, to see what it is, who looks at all particulars, and hands an exact report of all he sees up to Eventuality,
who makes a faithful record of the same, that Causality may use
it when it shall be needed.   Then here are Time, Locality, Size,
Form, Weight, Colour, and Order, living just around Eventuality, who make daily and hourly—yea, momentarily—reports of
the several particulars that belong to each to give of everything that passes under their notice.

Then on the inside of Causality stands Comparison, whose
office it is to draw analogies between the treasures of Eventuality and the conclusions of Causality, to make them clear,
make them *seen* by the whole perceptive group.   The perceptive
organs primarily know nothing but what they *see*.   It is the
business of Comparison to take the purely abstract deductions,
or the spiritual truths deduced by Causality, and compare them
with something the perceptives have *seen*, so that they can comprehend it.   On the outside of Causality stands Mirthfulness,
or Wit, whose office is the very opposite to Comparison's, viz. :
to show differences.   It takes the deductions of Causality, and
shows the perceptives wherein they differ from something they
have *seen;* and in showing these differences it often makes most
ludicrous pictures, throwing the whole family into convulsions
of laughter, from which circumstance it has been named Mirthfulness, or Wit.   Thus, in the frontal region, to guide and

direct the whole estate of the mind, is the intellectual family or group. Around this family is situated the semi-intellectual family, the constructive and imagining powers, used frequently by the intellect for its most grand and lofty purposes. Words can never express the beauty and harmony of this intellectual arrangement. Order, precision, utility, and perfection mark the whole of it. It is a beautiful and wonderful evidence of the incomprehensible skill and wisdom of the great intellectual Architect. How is it possible that such an arrangement could have come by chance, or without any previous design originating in perfect wisdom? At every action of our intellectual powers we involuntarily make an unanswerable argument for the existence and perpetual rule of a God of infinite skill and wisdom.

On the top of the head, as though to be the crown, king, and glory of man, and joining estates with the intellectual group, is the moral association, that galaxy of celestial inhabitants, that family of angels in the city of the human soul.

The centre and ruler of this group is Veneration, the reverent worshipper of God, the high priest of the church mental. Around him are gathered his family of celestials, robed in their garments of white. Immediately in front is Benevolence, the good Samaritan who blesses with prodigal hand all the children of need, and reports to Veneration that God's children are made happy, that Veneration may praise God for this grand result. Benevolence, too, lies close to the intellectual association, and can get any advice needed on its errand of charity in a moment. On each side, and between Veneration and Benevolence, is located Spirituality, the great seer and prophet of the soul, which points out man's spiritual relations, and opens the vista of future and immortal being. Back of Spirituality, and on either side of Veneration, is Hope, "the anchor of the soul," the great inspirer and stimulator to the attainment of good. This gives to Veneration a thousand pæans of thankfulness to offer to the great Father. Back of Hope is situated the ever faithful lover of right, and preacher of duty and holiness named Conscientiousness. It breathes through

Veneration its perpetual prayer for the triumph of principle. Back of Veneration and above Conscientiousness stands Firmness, holding continually the helm of the human will, and preaching stability to the entire family which lives and labours below.

The moral beauty and magnificence of this heavenly group is past all description. Each member is a legate of God, preaching the virtues and duties that belong to man as a moral and accountable being, an heir of immortal destiny, a member of the universal family, a kindred of angels, a being of magnificent powers of will and wisdom. While this family of celestials dwells in the mental world, it is vain to say there is no God, no religion, no heaven, no spiritual world; for its members are spiritual witnesses of these great truths.

Back of this group is found the family of selfish sentiments, which are ever consulting the dignity, importance, and nobility of this wonderful child of God, *I ;* its office is to make due provision for the attainment of whatever will promote its true excellence and glory. First is Self-Esteem, the preacher of human dignity; the second is Approbativeness, the lover of glory, or the applause of men—the inspirer of ambition. It lies on each side of Self-Esteem, which is located back of Firmness. Below these, and in the back, or occipital region, is located the family of lovers. They live for naught but love. The atmosphere they breathe is love ; the food they eat is love ; love is the light that cheers them and the fire that warms them into activity.

The centre of this family of affectionate principles is Philoprogenitiveness, the love of offspring, of helpless infancy.

Below it dwells sexual love, the primary object and end of which is the production of offspring, the reproduction of the image of the Eternal One, the multiplication of intelligent beings.

On either side of Philoprogenitiveness dwells hymeneal love, an ardent, faithful friend of its one single object of devotion; proclaiming ever to the world, the beauty, utility, necessity, and joy of the matrimonial alliance and life.

Above this lives that ardent, clinging, vine-like being, Adhesiveness, the eloquent expounder of fraternal love, and faithful devotee of friends. Above Philoprogenitiveness stands the old homestead—the beautiful, the sweet, grey old homestead, rich with a thousand golden associations, thronging with memories of olden love and life, written all over with the stories of the past, and sounding with the sweet music of all the home voices, harmonious as the strain of angels, and ravishing as the full note of love.

At the base of the brain lies the group of the animal propensities, giving life, vivacity, courage, energy, point, to whatever is necessary to man as a physical being, having *personal* rights and landmarks. The *position* which this group occupies being the lowest, indicates that its members are to be subjects, servants, not masters. Their office is menial service. They are excellent servants, but ruinous masters. The position of the *moral group* being the highest, seated upon the throne, shows that it was made to rule, that its office is to rule over the whole or universal family. It is chosen of God to be president of the mental republic. Its laws, principles, teachings, spirit, must be obeyed by every member of the grand union, or lawless anarchy and consequent unhappiness will prevail.

The position of the intellectual group shows most clearly that its office is to lead, to point out the way, to pioneer, to remove impediments, to open a grand highway, to pave it with truth and over-arch it with light, in which the grand army of the soul shall march up to the heavenly gates, prepared to enter into the fields of universal harmony, where every tree and shrub is loaded with the golden fruit of perfection. While the position of the family of lovers, being back, clearly indicates that they shall avoid the public gaze, and enjoy in sweet retirement that faithful friendship, those fond embraces, and dulcet pleasures, which they alone know how to give, receive, and appreciate.

In the examination of the head, the first thing to be observed is, the relative strength of these several groups. To understand the strength we must observe the length of the organs, or the distance from the centre of the brain, which is very nearly between the external opening of the ears. We must then observe the comparative size of each group or the amount of surface which each group presents toward the skull. From these, the grand characteristics of the mind may generally be determined with great accuracy. These can usually be estimated with approximate correctness by *looking* at the head. If the base of the head is wide and deep, the animal group is strong. If the front of the head is wide and long, the intellectual group is correspondingly energetic. If the top of the head is wide and high, the moral group is powerful. If the back head is large and full, the affections are full of ardour and strength. These several groups are but associated communities in the mental republic. When all are united, they constitute a mental unity. Hence, they exert reciprocal influences over each other. Hence, if the affectionate region is strong, with moderate intellect, the intellect will be made the servant of the affections; its highest feats will be performed, its noblest efforts put forth, when stimulated by the combined power of the loves; its judgment will be controlled, and its actions modified by the influence which is thrown around it by the pleading voice of the affections. If the intellect is strong with strong affections, then they will exert a mutual influence over each other: the intellect will guide the affections while the affections will empower the intellect. If the moral is very strong, with moderate intellect and affections, the moral will lead and adorn the character; but the moral will lack the power of the intellect and affections, to make manifest its lofty energies. Unite with it strong intellect, and the intellect will then become its counsellor, adviser, teacher, and the energies of the two combined will greatly augment the moral as well as intellectual power. Add to these, strong affections, and the whole character is made more powerful and grand. Thus the several groups work for, assist, empower, aggrandise each other; and the character is perfect only when the several groups are equally powerful and harmoniously combined. When this

union is complete, their powers mutually joined, their best
action secured, is true human grandeur and happiness attained.
Too much attention cannot well be devoted to the mutual influence of the several groups of faculties upon each other. For
it is thus we learn to read the characters of others and to improve and perfect our own.

This thought should ever be an inspiring one with us all, *the
improvement and perfection of our own characters.* For this we
should study this and all other sciences ; for this we should live,
love, adore, and think ; for this we should labour, strive, and
pray. The glory of our characters, the grandeur of our actions,
the splendour of our achievements, consists in living with this
as the quickening aim and object of all our lives. How glorious
is the life of youth devoted to self-improvement ! The love
of excellence, the love of progress, the love of perfection, how
beautiful when it burns a living flame in the heart of the
young ! Around youth's brow it weaves a wreath of glory ;
along his pathway it sheds the dewy nectar of life ; and into
his soul it pours the living spirit of mental beauty. Oh, God
of love, grant to all youth this heaven-born aspiration !

AFFECTIONATE GROUP.

In the examination of the several groups, it is proper that
we should commence with the affectionate—the group of lovers ;
indicated by length and breadth of backhead.

It might be remarked here that every faculty of the mind is
an affection. We talk of the intellectual, moral, and social
faculties, as though they were different in their natures, as
though the social was a *love,* while the intellectual was a *thought.*
What is the difference between a love and a thought ? One is
the offspring of a *social* faculty while the other is the offspring
of an *intellectual* faculty. They differ only in the object which
called them into being. They were both conceived in affection
and brought forth as the legitimate offspring, each of the particular faculty that gave it being. What then is the difference?
In the mental act which calls them into being is there any ? I
believe there is not. Every faculty is really a *love* or a loving

power. But each one has a different object. Toward that
object each feels alike, thinks alike, acts alike. Some are
objects of life, others are objects of principle. Adhesiveness
loves friends ; Self-Esteem loves self; Veneration loves God ;
Conscientiousness loves truth, right, holiness ; Hope loves a
glorious future ; Benevolence loves an object of need ; Ideality
loves beauty ; Comparison·loves analogies ; Wit loves differences, incongruities ; Causality loves the relations of cause and
effect ; Acquisitiveness loves money ; Constructiveness loves
mechanics ; Tune loves music ; and so on to the end of the
chapter. Each faculty loves its object—loves it with a deep,
warm, ardent, faithful affection. And that affection, the interest
which each one has for its object, is a *love*. Some of their
affections we name thoughts, some affections, some aspirations, some passions, but really they are all loves, in the true
sense of the term. Man's whole active nature is expressed by
the word *love*. The only difference in the different faculties is
in the objects upon which they fix their affections. Thus man
is capacitated to love everything that God loves. And when
he does, when every faculty of his mind is fully and perfectly
gratified by fixing its energies upon its particular object, then
will its happiness be complete, and its glorified state attained.
But is there no difference between what we call the affectionate
group and the other portions of the mind ? If so, what is their
difference ? The affectionate or social group consists of those
faculties which fix their attentions upon the different classes of
the human kind, such as sex, child, friend, companion. They
are those faculties which induce man to associate in some way
with his kind, which bind him to his fellow in the varied relations of social life. Hence all the associations, alliances, com-
pacts formed among men, have their origin in this group. The
family, the religious, and the political associations, are all formed
at the call of these faculties. Hence this group is the social
group. Here the principles of Association, which are stirring
the world so powerfully, have their origin and support. But
we have not time to generalize; so will examine each faculty
by itself.

AMATIVENESS.

The first is Amativeness, the primary office of which is sexual love. It is the grand bond of society, the bottom principle of the great social confederacy, the mainspring and moving power of human life, developments, progress, and happiness. It is a pure, a grand, a noble affection; as worthy of respect as any implanted in the human soul. All true men and true women respect it, admire it, as devotedly as they do the spirit of charity, or the worship of God. It constitutes a part of the human soul, and is not less holy and noble than any other. Its use, utility, and office are worthy of our devoutest meditation, and intensest study. There is, perhaps, no love more tender, more earnest, more self-sacrificing, more faithful, enduring, and deep than this; none that enters more largely into human welfare and happiness, administers more to human virtue and refinement, and works more powerfully upon human destiny. The primary office of this faculty, as connected with earth, is the reproduction of the species. But its grand, final, eternal office is to bind the two great halves of humanity in one great and golden whole by the filaments of a love as deep and deathless as the nature of mind. Viewed in this light, it is a sentiment lofty and pure beyond all powers of expression. It is designed to be the hymeneal link between man and woman through endless ages; uniting their powers, inspiring their energies, strengthening their virtues, magnifying their natures, ennobling and glorifying their whole souls. Its end is to breathe a holy rapture into life, spread a serener, yet inspiring charm over our whole being, and awaken the noblest emotions of unselfish affection that created beings ever can feel. Its use, as a stimulant to action, as an inspiration to virtue, is and always will be, of incomparable value. Neither man nor woman standing alone, is perfect in the action of their minds, nor can be, till this love has bound them into one, electrified their souls with the lightning flashes of its holy sentiment. It paves the way for other loves as strong and pure as this, and implants one affection after another as its legitimate offspring, till the whole domestic group is pouring out its tides of fervent and varied love. To this the domestic loves owe their origin. For their quickening energies they are indebted to this. This is, so to speak, the parent of all love.

Man without this love is cold, reserved, severe, coarse, vulgar, and debased. Refinement to him is an idle name. Affection another name for selfishness; seldom ambitious of good or great things, uninspired by the tenderest voices that whisper in the inner court. He is like a barren tree in the desert. About him no green thing flourishes: and around him gather no human beings for succour and support. He is at best but half a man in power, character, and influence.

Similar remarks might be made of woman unawakened, uninspired by the magnetic charm of this holy affection. Her most beautiful character, her most charming influences and power, her most angelic spirit and devotion, are given her by the radiant flames of this kindling altar-fire. It is in vain to expect woman to appear in her highest, noblest, purest character till her whole soul has been quickened and kindled into a flame by the stirring impulses of this great passion of the human heart. This more than doubles the native charms of her character, and greatly augments her power of mind and heart. Man and woman were formed that each might be an inspiration to the other, and it is in this sentiment that the inspiration is enkindled. They are for each other the object of the intensest affection. Out of this affection many of their purest joys grow. It is the source of the tenderest and sweetest delights of life. The most ravishing charms of being have their origin in the elevated action of this affection. When its love is consecrated by the hymeneal rite, sanctioned by the moral sense, and guided by intellect, it is a pure, spiritual devotion, and is the spring-source of an elevated and perpetual ravishment of soul, delightful and holy as the loftiest virtue. Yea, it is a virtue high and holy, a virtue binding upon all men and all women to exhibit, a virtue that is the parent of many others, and that opens a world of tender and precious delight.

I know very well that this sentiment is, and has been, more abused, perhaps, than any other. And this very fact shows

that it was designed for, and is capable of, conferring the greatest and purest of pleasures. It has been abused because of its
wonderful charms. And its abuse is followed by more wretchedness, degradation, and utter damnation than the abuse of any
other. This, too, is proof of its excellency and power. One
great reason of its general abuse is found in the ignorance
which is almost everywhere prevalent concerning its true
nature and office, and the laws by which it should be governed.
It is subject to fixed and immutable laws. When these laws
are obeyed, its joys are complete and rapturous; when they are
disobeyed, its miseries are sure and deep, its degradation black
and foul, and the ruin it works indescribably awful. It is the
hot-bed of all vices, the grave of all virtue, the death of all
happiness. It has overspread the world with its wrecks of
ruin, and planted its cankering thorn in ten thousand wretched
hearts. And in many instances it has done this under the veil
of that terrible ignorance which overshadows the minds of men
concerning the laws which govern this powerful affection, and
the great end and object it was designed to work out. In only
a few instances, probably, does it perform its full and blessed
work. In the present state of enlightenment it is almost everywhere subject to the most awful and degrading abuses; and
the vast majority of those enjoying its privileges are reaping its
harvest of miseries. Many are withering under its blasting
flames without knowing even the source of that wretchedness
which they feel is eating out the life of all their joys. The
statute-book of this almost omnific love has been a sealed book.
The science of its government has been unstudied and untaught.
Oblivion has covered the lives of those who have reaped its
highest joys, as well as those who have been withered by its
stinging miseries. The history of its virtues and its vices have
been only hinted at in dark disguise.

Foul insinuation, low joke, lewd allusion, sly innuendo, low
ribaldry, shameful slander, coarse distrust, and lying hints have
been the only language in which men have spoken of this strong
power. One who possessed it not might be in our world a long
time and not learn for a certainty that such a power existed.

He would hear hints that there was a spring-source of joy and
ruin somewhere; but where and what he could not learn from
history, science, church, or school, unless he stumbled upon
Phrenology, which the world has done its best to discard. To
me this concealment of the knowledge, principles, and laws of
this affection is a blighting curse, under which humanity is laid
by its wickedness. And the light and irreverent manner in
which the world treats and speaks of this subject is a deep in-
sult to virtuous principle, and a base slander upon the purity
and excellency of this affection. Whenever I hear the sly and
lewd hints, jokes, innuendos, and puns about this affection or
the marriage relation and alliance that grows out of it, which
are as thick and as destructive of virtue in every community as
were the locusts of Egypt, I feel sure at once that they come
forth as the legitimate fruits of abused Amativeness. I set the
seal of condemnation upon all such persons as being destitute
of virtue.

The first and fundamental law of this affection is the law of
marriage. It chooses a single object, fixes on that its deep
regards, lavishes on it its warm treasures, and is true and
faithful. Of itself it would no more, wish no more. It would
protect, cherish, assist, and love that object till its last earthly
sand had run out. It would hold with it a sweet and perpetual feast of pleasure, not of carnal joys, but of spiritual
communion.

Thus, it would wed itself to its one object, and live with it a
life of serene peacefulness and pleasure. This is its primary
law. In obedience to this law it reaps its golden joys and
exerts its benign influences. In obedience to this it opens its
treasures of purity and affection, and spreads its rich feasts of
cheerfulness and peace. In obedience to this it stimulates to
noble actions and glorious achievements. In obedience to this
it becomes a grand principle of self-sacrifice, often rising to the
sublimest heights of virtue, and affording the purest and loftiest
pleasures.

But in disobedience to this it turns to the fire of hell, and
burns to the soul's very centre, consuming all virtuous prin-

ciple, eating out its peace, corroding its heart, and spreading the deadly upas of ruin through all the faculties and all the life. In disobedience to the law of marriage, it is the vice of all vices, the curse of all curses, the ruin of all ruins. Its flame is red with ruin and black with pollution. Its joys, and blessings, and virtues, are known only in the kingdom of marriage; and he who attempts to know them elsewhere is a traitor to God and man, to principle and duty, and is fit only for the fires of hell and the essence of the gall of bitterness.

Its second law is the law of purity. It is heaven's own law, and must be obeyed, or wretchedness follows with lightning haste. It must be obeyed in the matrimonial state, or it will work its ruin there as well as elsewhere. I ought to spend half an hour upon this law of purity, but time will not permit.

The organ of this affection is situated in the base or back of of the brain, and is called the cerebellum. It is separated from the cerebrum, or main brain, by a strong membrane, but is connected at the centre as all other organs. Its material is the same as the rest of the brain; and it is covered with convolutions even more densely than the rest, showing that it has more power in proportion to its size. In man it constitutes one fifth of the entire brain. In woman it constitutes one eighth of the entire brain. This shows the mighty power that it exerts in character, and the importance of our possessing a perfect knowledge of its end, use, action, and laws.

NOTE.—All phrenologists agree with the author in attributing great importance to a correct knowledge of the organ of Amativeness, and how to use it aright. A special class of literature on the subject has been developed, to which the reader is referred for full information. Send to the Publisher for the SPECIAL LIST.

LECTURE V.

Parental Love, its Office and Necessity—Reason, Conscience, and Benevo-
lence no Substitute—Sacredness of the Mother's Love—Parental Love
Unselfish—Other Faculties acting with this—Anecdote of Parental Grief
—Children, sources of Parental Happiness or Misery—Influence of
Parental Love on Childhood—Abuses of Parental Love—Adhesiveness—
Society founded on Adhesiveness—Civilisation and Power the result of
Fraternity—Fourier and the Quakers—Christianity a Fraternal Spirit—
Friendship the Charm of Being—Solitude—Home-sickness—Adhesive-
ness an Element of Success — Its Abuses — Inhabitiveness — "Sweet
Home"—Charms of Home—Man a Local and Social Being—Evils of
Scattering a Family—Every Family should have a Home.

PHILOPROGENITIVENESS.

WE again invite attention to the social feelings. The love of
offspring will first claim our attention. In phrenological science
it is called Philoprogenitiveness. It is the next-door neighbour
to sexual love, and seems very naturally to grow out of it. The
ultimatum of sexual love, in our present mode of being, is to
produce offspring. They must be cared for or they would
perish. Ordinary friendship would not care for them, for that
fixes its interest upon objects that can return its favours. It
wants, and must have, reciprocity of feeling and action. This
it cannot get from helpless infancy. Reason would not care
for them. That might point out the ways and means by which
they might be protected and sustained; but it could never,
would never, nurse, cherish, and tenderly guard them. It
would never shield them from the storm, nor answer their
thousand little necessities. Many men of powerful reason
exhibit little or no interest in children; and many others of
very inferior intellects have burning affections for their children.
Benevolence would not care for them sufficiently to watch over
them day after day and year after year till they should come to
maturity. This would cherish them long enough to supply their
present wants; then it would leave them for other objects.

Conscientiousness would not prompt to that perpetual tenderness for which their helplessness calls. No moral principle will inspire that sacred and sensitive regard for them which is necessary to rear them to manhood. For often do men and women, of the purest and loftiest moral characters, exhibit but little interest in children, while, on the other hand, individuals destitute of all morality sometimes exhibit the most passionate fondness for their own and others' offspring in infancy.

The conclusion is irresistible that there is in the human mind a separate and distinct faculty, the sole office of which is to inspire a true, a faithful love for its offspring.

The daily evidences of this are seen on every hand. Witness the mother's watchful care and vigilant tenderness. She is the guardian angel of her babe. The first sight of it gives her a wild inspiration of joy. She gazes upon it in its utter helplessness as upon the concentrated treasures of a thousand worlds; its very breath inspires raptures in her bosom. From her heart there leap a thousand angel-prayers for its welfare. Through her spirit an unearthly tenderness breathes, like the spirit-utterings of angel-hearts. All absorbed in her one tender thought, she proves that all the mother is made by God. I have often thought that if God has any representative on earth, any type of Himself, any living, breathing image, if He has given form in this world to an idea of the next, given us even a shadow of heaven, it is found in the mother.

Who can doubt that the mother, all the mother, is God's own work? And who that has had and known a mother, and felt her love, can doubt that her great Author is love, pure, passionless love? The best evidence that God is love, is love in its proper form and tenderness, is love in its immaculate, unselfish glory, is love in its cherishing blessing, sweetness, and majesty, is found in His best work on earth—the mother's heart. I speak of the mother's instead of the father's love, because hers is usually stronger than his; but sometimes it burns with equal strength and ardour in the father's heart. It is *parental love.*

When it is found existing in all its strength and ardour in both father and mother, what a heaven of delicious sweetness is poured around the little helpless mortal which they call their babe. The very atmosphere it breathes is loaded with the spirit-fragrance of their hearts, that are now blossoming with this rich flower of the hearthstone garden.

> " Watch the growth of this *babe ;*
> See how its life is guarded,"

see how its wants are anticipated, see how its happiness is consulted, see how its couch is smoothed, its slumbers watched, its pathway decked with flowers, its nourishment supplied, its whole being made sensible of the perpetual presence of its guardian love.

In sickness and in health, in gladness and in sorrow, in virtue and in vice, in obedience and in sin, it is still the object of a deathless parental affection. Through all the varied vicissitudes of fortune it follows him. The older he grows, the deeper and richer it flows, till it usually becomes the one grand, all-absorbing feeling of aged parents. At its altar is laid their richest sacrifices, and poured their fullest prayers. Often have I seen it rise to a majestic height, and exhibit all the glory and grandeur of the genius of love. Strong and wonderful is its power. No labour is too severe, no sacrifice is too great, no trial too forbidding for it to make. It forgets self, as though self existed not. It forgets all things but its own dear objects. When these come to maturity, and are able to reciprocate their love, and become companions for their parents, this affection unites with it the ardour of Adhesiveness, the clinging devotedness of friendship. Then its flames become richer and deeper than before. It becomes more absorbing than ever. When it is united with large Approbativeness, it is ambitious for its children. It covets for them the praise of men. It is willing to afford for them every means in its power to secure the trumpet of fame, to blow a note that shall give utterance to their names.

If Hope is also large, it paints for them a glowing future, one sparkling with the wreaths of honour and achievements that shall crown them with the laurels of universal esteem. If it is united with large moral organs, it will covet most the virtues of truth and righteousness for them. Its chief care will be devoted to the cultivation of their moral natures. Its prayers will always be breathed for a close and faithful walk with God. And if with this combination, Cautiousness is largely developed, it will be harassed with a thousand fears lest they may not ornament their characters with gems of a living, exalted virtue. There is, perhaps, no combination that suffers so much for its offspring as this.

The thousand little aberrations of children and youth from the path of rectitude, their foibles and follies, make their parents, with this combination, most intensely miserable. A little world of this misery have I already seen in my short life. I am almost a daily witness of it. Oh! if children knew how much their parents suffer for every act of impropriety and immorality they commit, methinks they would consider better their ways.

Permit me to relate one instance that fell under my notice. It occurred in a family where I was boarding five years ago. The family consisted of a husband, wife, and two children: a little girl of eight years, bright and promising as often blesses an earthly home, and a little boy every way her equal, but two years younger. I entered one evening, between sunset and dark, and found the mother in a most violent outburst of grief. She seemed bordering upon insanity. No words can express the intensity of her suffering. I was alarmed, and as soon as I could calm her enough to understand my intentions, I asked her the cause of her trouble. Her husband was near. He was nearly as much afflicted as she. The little boy was at a window, and the girl not to be seen. As soon as she was able, she told me, in broken accents, that her little girl, in whom the best of her life was centred, had been out at play among her mates, and had come in and told her a *story which she knew to be false.* She was shocked almost to despair. The thought that her daughter would not make a woman of truth, wrung the chords of her life. She had punished her as much as she was able. This was the second time she had caught the little girl falsifying. And now the awful thought that she was not to be trusted was breaking the mother's heart. It was long

F

ere I could console her, or fill her with better hope. It was a sad time for both the mother and the daughter. The work of that unhappy moment embittered several months of both their lives. I saw the family not many months since, and the little girl told me with a tearful eye, yet a happy heart, that she had never since been guilty of the least equivocation from truth ; and the mother added that, no mother was blessed with a more truthful and dutiful daughter.

I have often thought since how easily that child could have turned that young mother's whole life into one dark, perpetual scene of unmingled wretchedness. Probably there was no other way in which she could have been so severely afflicted. Parents suffer more from their children's ingratitude and immorality than from any other source ; more, perhaps, than from all other sources put together. This love of theirs is so deathless and tender, that, when united with high moral virtues, it becomes the source of their most intense happiness, or most excruciating misery.

I once heard a mother, who had reared a large family, say, with a depth of honest pride and joy that words pretend not to speak, that she had never known one of her children speak a falsehood, or equivocate one hair's breadth from the truth. These considerations show how much of a parent's happiness is thrown into their children's hands. It is mostly at their care and keeping. And what a powerful stimulus to good and honourable actions this affords, or should afford, every youth. What base ingratitude must fill that youth's bosom who will fill a parent's heart with the barbed arrows of his folly and immorality, who will bring down his or her grey hairs in sorrow to the grave.

When Philoprogenitiveness is united with large intellectual powers, combined with large Approbativeness, it covets for its offspring a career of intellectual glory. Thus it will be seen that this love is greatly modified by its mental combinations ; and the influence which it has over children will always be such as its combination shall yield. The primary office of this affection seems to be the care and protection of helpless infancy. It loves helplessness. It delights in little children, and the smaller they are, the more it loves them. It regards them as charming little creatures. It sees in their mute actions and half-discovered smiles a glory that the world's best geniuses cannot match. It gazes into the face of the sleeping babe with a kind of rapture; and it can talk with children, make itself understood by them; yes, and this, too, without saying a word. A person strongly endowed with this faculty can always interest children, make them his firm friends. They will love him as naturally as the electric current runs along its conductor.

Have you not noticed that some persons get the affection of all the children they meet? The children trust them at sight, and love them as quickly. These are they who have Philoprogenitiveness large. It has its own modes of expression, so peculiar and mysterious that they cannot be described; yet every child will understand them as readily as the most familiar household words. It loves to fondle, caress, and play with children; can never see a pretty child without wanting to kiss it, and is always the child's friend, advocate, and protector. Hence all persons should have this faculty in its strength. It should be strong in teachers of children. No man or woman can govern children successfully without a strong endowment of this affection. In this lies the great secret of success with children. Writers for children, toy makers and sellers, are generally strong in this affection. It is stronger in women than in men. Hence woman can please, nurse, persuade, interest, and benefit children more than men. The affection is, in and of itself, a noble sentiment, and should be cherished and cultivated well. A woman without it is not a woman. She is destitute of one of the brightest ornaments of woman's character. She has no right to become a mother; no right to become a wife, except under special circumstances. A man without it is, at most, but a portion of a man.

This affection is liable to great abuses. It is of itself a blind love of children; and if not properly directed by intellect, and elevated by moral sentiment, it will indulge them in everything they desire, and prove the ruin of the very objects it wishes to benefit. Many a fond parent has suffered intensely from the imperfect enlightenment of this affection. It should be the study of every youth to enlighten, develop, and elevate this noble affection.

Its organ is located in the occipital region of the brain, just above Amativeness. When it is large, the head extends back from the ears a great distance, and back also from the neck. When it is small, the back of the head appears to rise almost perpendicularly with the neck. Its location and comparative size can easily be determined by a little practical observation.

ADHESIVENESS.

We come next to the faculty of Adhesiveness, the origin and fountain of friendship. It is one of the most beautiful adornments of human character, and administers greatly to human happiness. The very mention of the word friendship thrills many hearts with sweet delight.

Adhesiveness is, strictly speaking, the full, flowing fountain of friendship. Here originates all its tenderness; here enkindles all its fires; here swells all its floods of dulcet emotion. It is this faculty in animals which causes them to herd together. It is strong in all gregarious animals. It is this that gathers the fowls of the air into flocks, and the fishes of the sea into shoals, and men into communities. It is the gregarious instinct, and we must not say that it does not administer to the happiness of animals as well as man.

Would you see an exhibition of its joy-inspiring power in an animal, go away from a favourite dog, and after a few weeks' absence return. What dancing gladness he will exhibit! He will become half frantic with delight; he will almost laugh outright for very joy. It is the charm given him by delighted Adhesiveness. He feels precisely as you do, when the best friend in the world returns after a long absence. In man it is the spring-source of the associative principle. Hence it is Adhesiveness that forms societies, communities, nations. It is Adhesiveness that forms copartnerships in trade, business corporations of all kinds, societies of every description, asso ciations in all their multiform characteristics, states, nations, kingdoms. Strike out Adhesiveness from the human heart, and the ten thousand societies, companies, and associations would dissolve like the frost-work of morning at the day-king's approach, and pass away into their primary elements. Poll-books and roll-books, and constitutions and name-lists would all become useless, and society—society, beautiful as it is to us now, interesting and lovely as we regard it, excellent and grand as it appears—would pass away, and men would wander in solitude up and down the earth, each in search of a daily existence by himself.

Look about us—behold our institutions, noble, time-honoured blood-bought, brain-earned, and heart-consecrated monuments of associative civilisation. They stand among us thick as the stars in night's diadem, and far more brilliant. They are the landmarks to count human ages by. Into them is poured the light of sixty centuries. Around them gathers the consolidated wisdom of the past. They are the associated effulgence and glory of all human achievements. They are the milestones of man's progress up his heaven-ascending career. They are the record of all victories, the reward of the life-labours of millions of minds, the light-houses that all nations have built around the ocean of civilisation. The elements of their institutions are cemented by the strong bond of Adhesiveness. Break this, and they all dissolve in ruin. Without Adhesiveness they could never have been built. Without this, man is an isolated being; he works alone; he is a Napoleon on an Elba, a Selkirk on an ocean-rock. Isolated, man is weak; associated, man is powerful. When a nation of hearts beat together, what a pulse they make! It is like the tide of an ocean. When a kingdom of arms are bound in one, what a power they wield. It is like an earthquake throe. When a race of intellects are digging at the mines of thought, what precious gems do they bring to light. And when those gems are all set in one crown, what a galaxy of glory do they present. And what were virtue, or talent, cr loveliness to the hermit?

It is Adhesiveness that makes a nation's heart beat with

one pulse, that binds together a kingdom of arms, that gathers in one blaze of glory the lights of all minds. Some people censure Fourier for his plan of social organisation. His idea was to perfect the associative action of men, to make their interest and their pleasure one, to pour through the whole fabric of society the cement of Adhesiveness, so that every beam and brace in the great frame should be thoroughly cemented together; yea, so that every pillar in the great temple, all its finish-work, even to its glorious dome that lifts toward the blue sky, should be sweetly and firmly wedded, each to its proper place, by the attractive power of Adhesiveness. In him the principle was strong and intensely active, and with a powerful mind he *evolved* a vast plan of association, that ages hence will be better understood and appreciated than now.

He saw that man's interest and friendship ought not and need not war with each other; that what friendship asks, interest grants; that all duties, when illuminated by the light of a universal friendship, become our highest pleasure.

He was a friend, eminently a friend, for it was by the principle of friendship, or Adhesiveness, that he sought to bind his vast machinery of hearts and souls together. It was this principle that originally inspired the Quakers. Mother Lee, no doubt, was strongly endowed with the affection of Adhesiveness. Hence her religion was a religion of friendship; her followers were *Friends*. They felt the workings of this sweet spirit-charm, they yielded to it, they named themselves for it, and their children for it, and their sect for it. *Friends* was the significant title by which they were known. To its spirit they devoted themselves. By its spirit they became remarkable for its kindly exhibition. The world has honoured, and justly honoured, the Quakers for their faithful practice of the spirit of Adhesiveness. It is in this faculty or affection that the feeling of fraternity originates. Hence the *brother* is found here. Here he puts his arms around his brother, and clasps him warmly to his heart. The feeling of brotherhood is first felt toward those of our own household or family. It puts out its tendrils, and binds them closely to us. The vine of brotherhood grows around brothers and sisters. It plants its roots in the soil of home. It gathers its nourishment from the crumbs that fall around the home-table. It winds its tendrils first around the inmates of the dear old paternal roof. It next reaches out to early associates and more distant relatives, and winds them into the folds of brotherhood. It next extends its arms to acquaintances, and next to their friends and acquaintances, till at length it reaches its embracing tendrils around the entire race. It will be seen, then, that the germ of the religion of the gospels is planted in this feeling. Without this, religion would be only a fervent aspiration, a perpetual prayer. This is religion's handmaid.

Religion without brotherhood would fail in its practical result; religion without friendship would be a glorious aspiration, but a sad and meagre work. There is, perhaps, no affection that works a more enrapturing charm than does this. Where it is strong, it is an overpowering feeling of love; it makes friends true as steel, faithful as the sun, and as enduring as the mountain rock. How beautiful does the altar-fire of friendship burn in some hearts, and how sweet is its holy incense!

What will not friendship do for its object? It will stem the mountain torrent, the winter's cold, the summer's heat, the storm-god's rage, and brave the ocean's perils. Days and nights it will labour in devotion and hope. Through evil and through good report it will burn on, the same steady flame. It affords to the labourer, to the scholar, to the professional man, one of the strong incentives to persevering effort. Weak should we all be, were it not for the influence which our friends have upon us. We all live as much for others as for ourselves. The highest charms of our being come through our friends. Think of our social pleasures, our flows of soul and feasts of reason; our sweet, entrancing seasons of joy. Their deepest ravishment of delight comes through this feeling. Take a walk, on a bright summer's night, with the friend you love best. Feel you a charm winding itself around your whole being, and lifting you into a sort of ethereal paradise? It is the genius of Adhesiveness breathing its inspiration through your heart.

Open a letter from an old and faithful friend; *read* its
burning words, running over with the brimming floods of affection. How feel you? Can you tell? What is that that quivers
along every nerve of your being, and trembles in spirit-echoes
through your whole soul? Ah, it is the entrancing spirit of
Adhesiveness. Go and live in some lone solitude, with no
human being around you; and what are your feelings? Is the
sun as warm there as amid your friends? Are the stars as
bright, the night as glorious, the flowers as beautiful, the air as
balmy, nature as delicious? What spreads a cloud over all,
and glooms your life in darkness? It is wounded Adhesiveness. Depart from all the friends you love, and go into a
distant land or place to dwell among strangers; where stranger-
tongues greet you, and strange eyes scan you; where no
familiar voice, sight, or sound, or face, cheers you; what is that
sad, sickening feeling which you experience, which you sometimes call " home-sickness?" It is injured Adhesiveness.
When you part with dear ones, why does your soul writhe in
agony? It is the bleeding of the severed bond of Adhesiveness.

It is natural for Adhesiveness to make friends; it gathers
those of a like precious feeling around itself; it gains their
confidence, and secures their friendship; it works a sort of
inspiration over those in whom the feeling is strong, and opens
a road direct to their hearts. We sometimes meet with persons
into whose hearts we can walk as freely as into our own parlour,
and to whom our own hearts are equally open; they are those
who have strong Adhesiveness. Strong Adhesiveness is quite
essential to success in almost all kinds of business; it secures
customers for the merchant, clients for the lawyer, patients for
the physician, patrons for the teacher, hearers for the preacher,
work for the mechanic, markets for the farmer, votes for the
politician. Every person who has it large has his particular
and tried friends—friends who will not forsake him, whom
money cannot buy, nor flattery seduce. Without friends, no
man can prosper in business; friends are his support, his
strength, his hope, his bond of success.

With the politician, the professional man, or any public character, strong Adhesiveness is absolutely necessary for success. This sentiment may be greatly abused; it may fix its affection on unworthy objects, or may open the heart to traitors; it may unveil its beauties to the deceiver and the hypocrite; it is very liable to be deceived; it is of itself a blind impulse of love : it is a poor judge of trustworthiness; it has a glorious heart, but no intellect; it is powerful as a giant, but unwise as an idiot; it is gentle as a lamb, but not wise as a serpent. Those who have it strongly developed cannot be too prudent. It requires great watching, or it will overrun all bounds of discretion; its outward language cannot well be mistaken; it is always fondling, caressing, handling its objects ; it loves to be near its friend, to draw him close to it; it has a most wonderful kissing propensity, and never gets tired of embraces ; it is all made up of tenderness, and seems to delight, above all things, in some expression of its deep sympathies with its object. It is very essential in a good companion, whether husband or wife. When united with large Amativeness, it gives inexpressible warmth, strength, and tenderness to the affections. When, with this combination, it is united with strong moral feelings—*Ideality*, and intellect, and a mental temperament—it is an elevated, pure, consecrated devotion to objects of kindred, elevation, and purity, so excellent and morally grand that language never has been, and never will be, trusted with its expression. It then rises to the sublimest poetry of the heart ; it is the soul's unwritten foresight of heaven.

The organ of Adhesiveness is located just above and outward from Philoprogenitiveness. When it is large, the head is wide through this organ ; when very large, the head is very wide and extended backward, giving a heavy lobe to the back part of the brain. It is easily distinguished, and its size can be determined with but little difficulty. It is usually much larger in women than in men ; hence they are the truest, warmest, and firmest friends, the most ardent lovers, and the most devoted companions.

INHABITIVENESS.

Next comes the old homestead, in Phrenology called Inhabitiveness. This is the home instinct, the home love. It consecrates that sacred spot, that clustering place of all the loves, known so long in song as "sweet, sweet home." It makes the place where we have lived, loved, and acted, the dearest, sweetest loveliest place in the world. To this faculty the sun is brighter, the rose is fresher, the water is clearer, the air is balmier, and nature lovelier about its home than anywhere else in the world.

In truth and verity, it adopts the sentiment, "*There is no place like home.*" That sentiment was written by Inhabitiveness. The whole poem to which it belongs was inspired by this faculty. This feeling never gets tired of home. It wants to stay at home, and stay at home, and keep staying at home, and the longer it stays, the sweeter grows home.

Visiting! it hates that. Travelling! that is the meanest of all pursuits ; the roamer is worse than a blackleg. If it starts on a journey, it gets homesick before it gets out of sight of home. If it stay over night away from home, it sleeps not a wink. If it is away from home at mealtime, it has no appetite. The sharpest wit ever put forth, away from home, will not move it. It sighs, and droops, and fades, when away, like a water-lily planted in a desert. It cannot, will not, live away from home. Home is such a charming place, so ravishing in beauty, so sweet in fragrance, so bright in sunshine, so environed in loveliness, that all else is dull, and dark, and stupid, in comparison with it. This is the way Inhabitiveness feels when it is very strong and active. Hence "home-sickness" has its origin in wounded Inhabitiveness. There is a "*friend*-sickness," that, in its suffering, is so similar to "home-sickness," that both generally go by one name. They are very unpleasant maladies, and the objects of them are worthy of our consideration. It is evident, at the first thought, that the object of Inhabitiveness is to fix man to one spot, to induce him to choose one place out from the broad earth, and build himself there a home, that there he may make happy his companion, there rear his children, entertain his friends, gather the good things of life, pursue the flowery paths of science, the gemmy walks of literature, the winding ways of philosophy; that there he may build an altar to his God, love and do good to his neighbour, found institutions of learning, charity, and religion, and do all those great and good things men can and desire to do, when they are fixed to one spot.

Inhabitiveness and Adhesiveness make man a local and a social being. His nature bids him associate, but it tells him to do it at home. It tells him to bind around home all the attractions, all the advantages which his desires demand. Home should be the pleasantest of all places. At home should be his associates, his companions in labour, in learning, in religion, in amusements, in love. At home should be his school, his library, his laboratory, and observatory. Home should be his sanctuary, his church, and everything which goes to make the man pure, learned, wise, and good. This is surely the teaching of Inhabitiveness and Adhesiveness. How can this be done, when men live in such isolation as they do now? How can homes be homes, in the present organisation of society?

As it is, home must daily be broken up; children must separate from their parents and from each other; friendships must be formed, only to be broken; homes established, only to go to ruin. Ere the child's mind is half developed, it must be sent away from home to be educated, away from the very place where it ought to be educated. A son or a daughter marries. Instead of bringing home a companion, a friend, to adorn, enrich, beautify, and enlarge the home circle, home love, and home joys, one is lost; and a shock, a terrible shock, given to all the home affections. A pillar has fallen from the temple of home; the first presage of ruin. Soon another goes, and then another. At last down comes the old temple. The aged pair who reared it in love, and labour, and hope, are cast out upon the world's frozen ocean, to end their days in homeless wretchedness, sick of life and courting death.

This is but a picture of every home. This very moment a million of homes are thus tottering to ruin in every country. Oh, what an awful abuse is this of our natures! God designed and fitted them for a better fate. He formed them for a glorious home on earth, where they could be born, live, and die, in the exercise and development of all their noble faculties. Oh, aged fathers, mothers, throughout the world, how ye are now suffering for the great and general transgression of the law of Association, written out in the faculties of Adhesiveness and Inhabitiveness! And brothers, sisters, friends, who are widely separated by this same transgression, how ye, too, are suffering! When will ye learn to associate at your homes, and thus by mutual assistance, build homes that shall be permanent and glorious; where shall be gathered comfort, learning, and religion; where companions shall cluster, beauty dwell, and civilisation, in its true glory, bear its happy children up the highway of eternal progress? I confess, my young friends, that my heart is sick, absolutely sick of the wrongs and outrages committed against poor human nature by our present crazy, sinful, social organisation, or rather disorganisation. Speak of reform, and men will hoot in their misery like the owls in their blindness. Talk of a social organisation, and they will cry selfishness; forgetting that it is the very want of organisation makes men so interestedly selfish. Let them learn the lessons taught by nature, which are God's own lessons; let them learn the philosophy of the human soul; let them acquaint themselves with anthropological science, and they will one day see the right, and pursue it. There is surely a good time coming, when right shall rule over might; when homes shall be permanent, and be real *homes*, instead of temporary staying-places. Inhabitiveness clearly teaches that every family should have a home. If God has given man a love of home, He has given him a right to a home. Hence the law of homestead exemption is a righteous law, and accords with our natures.

Again, I might remark, that patriotism—the love of country—grows out of Inhabitiveness. The workings of this beautiful, strong, and honourable affection, it would be pleasant to trace; but time forbids.

The organ of *Inhabitiveness* is located directly above *Philoprogenitiveness*, and between the two organs of *Adhesiveness*. It is under the occipital suture.

LECTURE VI.

CONCENTRATIVENESS.

PASSING from the organs of the home affections upward, we meet at once with Concentrativeness. This organ, by some phrenologists, has been considered as united with Inhabitiveness forming one organ, the office of which is to give fixedness of mental feeling, permanency of mental state. But most phrenologists regard it has a separate faculty, having a close similarity in its office to Inhabitiveness. Inhabitiveness fixes in the mind a love of place, binds man to one locality, prevents him from wandering, gives him permanency as a locomotive being. Concentrativeness does the same for man as an intellectual being. It gives permanency to the intellectual states, a distaste for mental changes, or changes of mental action, a love for stability of mind. Inhabitiveness would fix the body in one place, while Concentrativeness would fix the mind in one state. There is a close analogy between the offices of the two organs, as we should expect from their being located together. Concentrativeness gives fixedness to the attention. When the mind engages in any action it lends its energies to render that action permanent, to continue it until its object is attained. It wars against doing two things at a time—against dividing the mental energies between several objects. It was Concentrativeness that first gave utterance to that trite old saying: "He who has many irons in the fire will be sure to burn some."

Large Concentrativeness will permit the mind to do but one thing at a time, will enable it to give its whole attention to one subject till that subject is thoroughly exhausted. It concentrates the mental energies into a focus, bringing all the powers to bear upon one point.

The power of attention is invaluable in all mental pursuits. It is the grand secret of success. He who concentrates every energy of his mind upon any subject, penetrates that subject, grasps it, comprehends it, and makes it his own. When he abstracts his thoughts from everything else, forgets all but the one thing, and pours his concentrated powers upon that, as does the convex lens the rays of the sun, he becomes master of that one thing. Often more depends upon this concentrative ability than upon brilliant powers. One moment's pure, solid, close, abstract thought upon any subject is worth more than a whole week's wandering, desultory, inconstant thinking. The one burns into the subject, the other glances around it. The one snatches it with power, and masters it at once with a giant's strength; the other tugs away at it like an infant trying to move a mountain. The one sees it an absolute reality in the clear sunlight of perception and reason, the other gets only a dim outline of it in the mist and darkness of doubt and uncertainty.

The logician, the student, the artist, the musician, who has the power of attention strong, and who buries himself in his own thought, will exhibit a power, and win a success and a victory, that will scarcely form a part of the vagaries of the inconstant dreamer who wanders over the whole creation a dozen times every hour, which ought to be devoted to abstract thinking.

If man had no Concentrativeness, what a whirlwind of changeability would he be. Every mental faculty would be at work at a time, and each in its own way, every one clamouring for a different object. Singing, fighting, praying, loving, reasoning, travelling, staying at home, braving, fearing, grasping, giving, and a score or two more things would be going on at once. Such a medley of discordant views, opposing interests, and avocations as the mental workshop would present, no man hath even conceived. Concentrativeness is a sort of helmsman, directing them to a single port at a time, and steering directly to that till it is reached; and then turning to another and pushing for that till it is reached; and so on, doing one thing at a time till all is accomplished.

Large Concentrativeness is distracted with the jargon of several objects before the mind at a time. It can bear but one, will have but one, is made miserable by more than one, gets nervous, fidgety, and out of patience when it is disturbed, or any question is asked about anything else; it cannot live in confusion.

Small Concentrativeness can talk about twenty things in a minute, give its opinion on forty different subjects, cast a glance at forty different sights, hear as many different stories, tell as many, and turn sixty complete mental summersets every half-hour, and be perfectly at peace and composed under them all.

It is a lucky thing for a common-school teacher to have but a small development of this organ, when he has sometimes ten classes to hear in half an hour, a hundred questions to answer on about as many different kinds of studies, fifty roguish boys to watch, and not a less number of roguish girls, and attend to all the paraphernalia of a common-school room; he surely has to make short turns enough to set large Concentrativeness perfectly crazy, while small Concentrativeness would be quite at home in it all.

A lady once asked me why it was that she found so much more satisfaction in the society of gentlemen than she did in the society of ladies. I replied, that there were various causes for such a preference, and that among them Amativeness stood very prominent. She answered that that was not the cause of this preference in her case, for that organ was small in her head and not active in her character. She then remarked, that whenever she met a number of ladies together, they conversed upon such a multiplicity of subjects, varied the subject of conversation so frequently, chattered so much like a nest of mag-

pies, that it nearly distracted her, and rather than bear it sh
had often retired abruptly and sought to amuse herself in solitude. Gentlemen, she said, she had found more disposed to
exhaust one subject before they introduced another for consideration, and on this account she had found more congeniality
in their society. She had large Concentrativeness.

SELFISH SENTIMENTS.

We have thus far been treating of affections, or sentiments,
that had external objects in view, or something beyond self.
We now come to another class of sentiments which are different, not in their nature, but in the nature of the objects contemplated by them. They fix their affections upon *self*, they
are devoted exclusively, absolutely, to the good of self. They
are strictly selfish sentiments. They have no interest in the
well-being of anybody else. They fix all the warm energies of
their deathless love upon dear, darling self. As the mother
loves her child do they love self. As the husband cherishes his
wife do they cherish self. As friend is bound to friend by the
filaments of a deep and deathless love, are they bound to self.
As the lover of home clings with a strong and abiding attachment to the place where he has lived, and laboured, and loved,
so do they wind their tendrils around the object that is all the
world to them. For them there is but one object, that is self,
and that is dear above everything else—" the world and all," to
them. In view of these sentiments, we see that it is as natural
for man to love himself as it is to love his friends or any other
object. Self is one of the objects of natural affection. Self-
love, then, should be preserved, educated, cherished as sacredly
as any other affection. It is a part of the mind, a part of the
immortal principle, a part of the living, eternal being which is
God's child and bears His image. Honourable, useful, beautiful, and glorious then is self-love.

Man is a child of God, and is as worthy of his own affections
as he is of the love of his Creator or his fellow. To be weak
in self-love is truly a mental deformity. To fail to cherish self-
love is to fail in a strong and imperious duty, even a duty which

Jesus recognised as an eternal duty. He says, as the great formula of divine law, " Love thy neighbour as thyself." He recognises self-love as the grand standard by which fellow-love should be measured. Let not self-love, then, be branded as evil. In its proper use it is as right and righteous, and as well-pleasing to the Great Father, as any affection in our natures. It is liable to abuse as well as any other, and requires the same guards, guides, checks, and cautions that every affection needs when it is strong. Still it is a truth, that self-love is more likely to be neglected than any other. Perhaps mankind generally abuses self more than anybody else. Self is often neglected, abandoned, cheated out of its just dues, and imposed upon in the most unscrupulous manner. Who protects self as he ought from all the dangers to which morally accountable beings are exposed? Who educates self properly? Who develops all the talent and glorious energies of his soul? Who adorns his mind with all the imperishable embellishments of virtue and truth? Who harmonizes his powers, magnifies his abilities, consecrates himself to the good, the beautiful, and the true, as becomes a child of the good God and an heir of immortal progress and glory? The answer is at hand, " No one." Then self is not properly, not sufficiently loved. I would gladly treat upon the moral aspect of this subject at length, but time bids me speed on.

APPROBATIVENESS

first claims our attention as one of the family of selfish senti-ments. This claims for self the approbation of men. It considers self as connected with a race of kindred intelligences, and it would bind them all together by a mutual respect and esteem. It can never live alone, never dwell apart from its fellows. Its nourishment, its very joy, is all drawn from them. They surround it with the summer-heaven of gladness, or im-merse it in Tartarian darkness. It is strongly, wildly devoted to its fellows ; but it loves them not on account of their good-ness, wisdom, or virtue, but simply for their praises. These it must have, or it withers in a worse than cheerless solitude.

G

The applauses of men are sweet to it as the songs of angels. They charm it into a wild delirium of joy. They thrill it with perpetual delight. They fill its cup of gladness to the very brim. Human applause is the grand object of its life. On this it feasts with a ravenous and insatiable appetite. Sweeter to it than the essence of honey is a full feast of praise. It thus affords one of the greatest stimulants to human exertion for whatever is great, good, or praiseworthy. To the scholar its voice is ever sounding in his ear, encouraging him to toil on amid every difficulty and danger, to spare not time, nor sleep, nor expense, nor ease, nor health, nor brain-sweat, for human hands will one day crown him with the laurel of a glorious and well-earned victory. It beholds for him, in the bright and opening future, a career of glory, and hence it bids him be cheerful and strong. Like a guardian god, it is always about him, whispering in his very soul its song of glory. To every man, in every business or profession, it comes with the same inspiring view of the glory that will attend him. And from the doer of good, from the cherisher of virtue, the blesser of the needy, the worshipper of God, it withholds not its inspiration. To them its voice is more subdued, its air more humble, its manner more in keeping with their several holy offices. But surely it fails not to attend them as constantly as it does the general on the battlefield, or the seeker of place and station. It always speaks of honour, distinction, glory. And its idea of glory is all found in the approbation and praise of others. There is, perhaps, no stimulus that is more universal and powerful, than that afforded by Approbativeness. Scarcely a human creature can be found unaffected by it. From the slave at his task in the burning sun, to the king on his throne, its rule is felt. Then, it is not only a stimulus to active exertions, to daring exploits, and almost superhuman achievements, but it prevents the commission of a world of crime, and the practice of as much vice. The hand lifted to do a deed of darkness and wrong is often stayed by the loud appeals of Approbativeness. Lusts are checked, passions curbed, slander's tongue disarmed, envy's work prevented, and the lawless career of

disobedience greatly narrowed by the stirring instigations of
this faculty. If in the field of its boundless ambition it lays
waste empires and makes nations groan in bondage, it at the
same time puts an end to a thousand old abuses of power,
breaks up a thousand haunts of iniquity, and deals a blow of
ruin to as many monsters in vice. It always does its great
works under the pretext of right, and generally believes that
great good is to be the grand result.

No faculty, perhaps, in the human mind is more liable to
abuse than this. When connected with great minds, unless it
is coupled with strong moral elements, it is the source of that
lawless ambition that overruns all bounds, that courts the whole
world for its sphere of action, that would sit upon the throne
of universal dominion, and be the one, only, all-grand, all-
imposing object of the adulation and praise of mankind. Such
it was in Alexander and Napoleon. Such men generally believe
themselves human gods sent for the deliverance and worship of
mankind. And under this delusive idea, given wholly by
Approbativeness, they often cause crime, devastation, and ruin
to overrun whole continents, and sow the seeds of a mighty
harvest-field of vice and wretchedness. So inspiring, so enrapturing is the voice of this syren in the soul, that they forget
all the laws of propriety, of right, of decency, and duty, and
give themselves up to its bewildering notes, charmed victims of
its single strain. Napoleon himself said, " Sweeter to me than
the voice of Josephine are the praises of the French people."

In lesser minds it is as often and as greatly abused. It
courts popularity; curries favour with the fortunate in worldly
matters; bows obsequiously to wealth and station; worships
equipage, dress, rank, fashion; conceals unpopular views;
affects to despise disapproved sentiments, although inwardly
known to be just; is also given to flattery, deceit, and often to
deep-toned hypocrisy. It induces its possessor to seek the ap-
proval of men, even at the expense of principle, duty, and
natural affection. But in all these abuses it utterly fails of its
object. Everybody sees the veil with which it attempts to cover
up its hollow pretensions. It is really the seat of vanity in all

its fantastic varieties of form, feature, and manner; and who fails to read "vanity" on all its silly works? A great world of poverty, wickedness, and wretchedness, the abuse of this organ causes. I would gladly descend to detail, and point out a thousand-and-one of its abuses, but I must not attempt the Herculean task.

Permit me to speak a moment of its natural language. It speaks out in its own peculiar way, and the real phrenological reader cannot fail to understand its well-written language. Its first and most significant sentence is this (a cant of the head sideways). This is its stereotyped speech, which it utters in everybody's eyes "from morn till night, from youth till hoary age." Its literal interpretation is "See here; don't I make a fascinating figure?" Speak a word of praise to a child, and see if his head does not drop to one side as quick as though his neck had been broken. Tell a milliner's lady that her bonnet is a charming thing, and lo! it will instantly hang on one shoulder, as though her neck had lost all its starch. Signify to a belle, just from a mantua-maker's shop, that her dress is a very Parisian beauty, and the bow in her neck will instantly resemble the graceful arch of the snowy swan, save that it will be turned to the side. Intimate to a poetess that her poetry is admirable, and lo! what a miracle you work with her head. It bows to the side like the top of a beautiful willow in a gentle breeze. Make a speech in praise of what anyone has said or done, and if he is in the assembly, he signifies at once that the hinges of his neck are well oiled. It is next to impossible for any man or woman to hold the head erect when under the influence of Approbativeness. Behold the bashful child, witness the diffident youth, see the blushing maiden. Not one of them can hold the head erect. See the gay belle when courting admirers; observe the youth when he would softly play the agreeable. The head of each will incline to the side, and not unfrequently wave to and fro like a reed in a breeze.

Besides this wave-like, sidewise motion to the head, Approbativeness gives a peculiar expression to the countenance. I is a soft, complaisant, gratified look, not altogether idiotic, nor quite intelligent; but a sort of self-gratified quiescence, the result of an inward pleasurable excitement, which shows itself in a half laugh, and a half grin, which no pen can describe, but which is perfectly visible and understandable to every scholar in Phrenology.

When Approbativeness is completely disciplined, it is one of the primary sources of genuine gracefulness of manner. When this is active, and united with active Ideality, it confers the peculiar charm of gracefulness, which is almost infinitely pleasing to everybody. It gives symmetry to all the motions of the body, harmony and apparently perfect naturalness to every gesture; ease to every action; flowing elegance to conversation; a ravishing sprightliness to the countenance, and all those sweet and flowing elements which combine to constitute the wonderful charm of gracefulness. Undisciplined Approbativeness makes one clownishly awkward; well-cultivated Approbativeness makes one charmingly graceful. This faculty confers that peculiar quality to the manners which men have named politeness. In all its multiform phases and characteristics it is the legitimate offspring of this love of approbation. Approbativeness loves to please, to gratify others, to play the agreeable, and hence makes its possessor desire to be sincerely and truly polite.

But I must not detain you on the thousand-and-one different outward appearances which this faculty presents. Its organ is located just above and outward from Concentrativeness, at the back and upper corners of the head. When it is large it gives width and prominence to this region.

Its locality is easily learned, and the mere tyro in the science can determine its relative size.

SELF-ESTEEM.

Next comes the pompous, magnificent aristocrat—the great everlasting " I "—the gentleman of splendid parts, of honour, of dignity, of kingly authority, in whom resides the prerogatives of sovereign power. His name is Self-Esteem. He is wonderfully satisfied with himself. He is a genius, and he knows it.

His judgment is superior to everybody's else, and he is sure of it. He is made of a little better material than any other human creature, put up in a more skilful manner, elaborated with greater precision, refined to a greater degree, and marked in flaming characters, "a superior specimen of humanity." His superiority is so apparent to himself, that it is a matter of very little concern with him what others think of him. He properly appreciates himself, he knows his own dignity, and feels his own immense importance, and that is just about as much as he cares for.

What are others to him? Mere lilliputian puppets, playing the second fiddle to him. He is lord; they are subjects. He is master; they are servants. He is first; they are second. What cares he for their opinion? It is not worth minding. He considers its source, and regards it as little as the idle wind that plays in dallying breezes about his temples. Other men are but flies, whisking in insignificance about him. They are very convenient, it is true, to do his bidding, and serve him in his wants; but then they are so mean in comparison with him that he cares not to commingle with their vulgar herd. Here and there is one formed of noble blood, wearing the stamp of true nobility. With those he can consort in high and honourable companionship. In their veins the blood-royal all flows.

See this gentleman of honour among his fellows. He walks with lordly mein. A calm, dignified self-complacency is written on the fixed and satisfied features of his face. His hat he supports as a crown on his head. His body he bears about as a precious thing. He robes it with care; feeds it on precious food; rests it on couches of superb comfort, for it holds the best drops of the royal blood. The ground, he treads on as though it was really *his* and not his Father's, and scarcely good enough for his footstool at that. If he speaks to another it is as though the king condescended to notice his subject. He looks down, and talks as though he did it just because it pleased *himself* so to do. He expresses his opinions as though they were absolute law. He despises all dissenters from them. They are fools who think differently from him. They ought not to be tolerated. He would crush them as puff-balls under his feet. He considers it bold presumption for one to be opposed to him, which ought to be immediately rebuked. His requests are all commands. His invitations are positive mandates. He loves to rule. The atmosphere about the throne is congenial to his feelings. Authority is the natural instinct of his character. He was born to be a leader. In whatever enterprise he engages, he must be first. No. 1 is marked on his brow. He is a man, and requires that all shall so regard him. He places an exalted value, not only upon himself, but upon everything that issues from himself. His labours are vast, and strikingly significant. The result of his efforts are but the elaborate products of genius. He reads his own writings, and is charmed by their elegance and beauty. The sentences are precise and clear; the periods lofty and grand; the thoughts bold and dignified. Few men can write with him. He listens to his own voice in public address. It is full and noble. Its modulations are the manly master-strokes of eloquence. What bold figures he uses! How striking! What fulness to all his periods; what power in his arguments; what vigour in his style of delivery! The same self-satisfaction marks all his actions, whatever be his calling or profession. He is a natural boaster; a constitutional braggart; an egotist from the centre outward. Everything that he sets his seal upon is a little better than anything else of its kind. If it is his, that is enough to make it better. He loves to talk of himself, of his wonderful exploits, his victories and achievements. He can never listen to the stories of others, because they always remind him of something far more important in his own history. You will observe that in his conversation he always uses the personal pronoun in the first person. "I," "My," "Me," are words of vast significance. If he has large Destructiveness and Combativeness he is a tyrant, an oppressor, exacting and severe. He loves to wield the sceptre. Authority he loves to exercise. He never tires of ruling.

This is Self-Esteem. Its abuse is tyranny, egotism, arrogance, pride, haughtiness, self-conceit, presumption, impudence, boasting. Its use is to give self-reliance, self-respect, dignity, confidence, a proper regard for our own rights, opinions, privileges, character, and standing as a child of God ; to impart a tone of real nobility and dignity to all our actions. This faculty should be cultivated in the young, for without it, man lacks that spirit of manliness, dignity, and honour which constitutes one of the main pillars of a reliable and virtuous character.

Its natural language is clear and unequivocal. It is the language of dignity. It bears the head high ; the body erect; gives fixedness to this natural position, and generally a slow, solemn movement to the whole body. I need not speak at length of its language. It cannot well be mistaken. Some of its actions are often confounded with Approbativeness. Care should be taken to avoid this.

The organ of this sentiment is located above Concentrativeness, and between the two organs of Approbativeness, at the crown of the head. When it is large, the crown of the head is high.

CAUTIOUSNESS.

Here is one more faculty that has a dear love for self and is frightened almost to death if self is in the least possible danger. It is the sentinel on the outer wall of the soul to give warning of the approach of danger.

Did you ever see a flock of crows alight in a corn-field, or upon a carrion ? You must have observed that one crow always places himself upon some high tower, or observatory, from which to look out for danger. Whenever he fancies that danger is near, he gives the signal and takes to his wings. In a moment the whole flock are making off as though death was on their track. This watch is the faculty to which we refer. Its name is Cautiousness. It is the sleepless soldier of the camp on the outpost. We are exposed to dangers on every hand. Enemies lurk in perpetual ambush about us. Disease floats in the wind, is coiled in our food, our drink, and rises in miasmas from the earth.

Death has his bow bent and his arrow aimed continually at us. Ruin is riding his red chariot on our track at the speed of the whirlwind. In the cloud is concealed a deadly archer, whose eye of fire is fixed upon us, and whose bolt of flame is quick as the glance of thought, and ruinous as the breath of destruction. Pitfalls are beneath our feet; floods are sweeping around us; pestilence walks at noonday, and steals in lurking silence at night. Our reputations, our fair characters, are as much exposed as are our lives. Surely we are in need of a faithful sentinel upon the highest eyrie of the soul to warn us of the approach of danger. The most vigilant and keen-sighted that can there be stationed cannot foresee all dangers and warn us against all harm. It is not in human foresight or power to know all the invisible approaches of the multitude of enemies that stand around man, thick as tombs in a grave-yard. But some may be seen. It is the office of Cautiousness to use the utmost vigilance as the sentinel of the mind, to watch for danger from every point, and to exhort every faculty to prudence, to consideration, to close circumspection. It is its office to hold a perpetual check on the hasty and turbulent impulses of the mind, and plead with them to "let their moderation be known unto all men."

Every faculty of the mind would run wild in excessive extravagance, were it not for Cautiousness to hold it in check. The passions and the appetites would know no bounds, the affections would be flames of unquenchable fire, the sentiments would know not but that they might clamour in passionate anxiety, day and night, for the objects of their desire, did Cautiousness not hold its steady rein, curbing their unbridled licentiousness. Man would not only bring himself to ruin by his reckless exposure to physical dangers, but he would ruin his mind by an excessive gratification of all the mental desires. He would burn up his soul by the flames that are enkindled within it, and which, held in check, constitute its glory and its grandeur.

Cautiousness may be regarded as the great regulator in the mind, holding every part in its proper action, and controlling all by its prudential dictates.

The mind is made up of hot impulses on the one hand, blind as stones, and clamorous as hyenas for their prey, and checks ,and balances on the other, to hold their power in proper subjection. Cautiousness may be considered the grand balance-wheel, influencing the whole by its steady and prudent movements. The beauty which this regulation affords is truly delightful to contemplate. Who can look for a moment at the mind and not exclaim, "How wonderful, how sublime! The evidence of a God is here." Long ago it was said, "An undevout astronomer is mad." With how much more emphasis may it be said, "An undevout phrenologist is mad." I am lost in wonder and amazement when I attempt to glance at the sublime excellency of the mental constitution, and the infinite wisdom and love displayed in its creation. Even in these bird's-eye views which we are now taking, we cannot fail to see astounding displays of creative skill and wisdom, and the clear evidences of equal goodness. Study these, my young friends, in humility and reverence, and give God the praise.

When the impulses are very strong, Cautiousness should be strong also, in order to hold them in proper subjection. When the impulses are weak, Cautiousness should be correspondingly weak. Some minds require large, while others require small Cautiousness. The effects which Cautiousness exhibit are to be determined, not by its size alone, but by its size as it relates to the strength of the impulses. In some, Cautiousness watches for physical danger, in others it watches the character. This is to be determined by the combinations.

Large Cautiousness prevents great risks, great exposures, and great efforts, and hence often makes a man a small man, who otherwise would be a great man. "Nothing ventured, nothing made," is the old saying, which large Cautiousness never approved. Thus it is clear that this faculty may exercise too great an influence in the character. Its suggestions should be thoroughly weighed by the intellect. To this it should always make its appeal.

Bashfulness, or diffidence, in children and adults, which is so painful to endure, and a source of so much awkwardness, arises from excessive Cautiousness. Care should be taken in the training of such persons, not to inflame this faculty by threats and by frightening them with real or imaginary dangers.

Excessive Cautiousness, with small Hope, produces melancholy. Those having such a combination should associate with those of an opposite cast of mind, whose influence would dispel their fears and throw sunshine on their pathway.

The organ of this sentiment, which is so watchful of self, is directly on a line between Approbativeness and the ear, joining domains with Approbativeness. When large it gives great width to that region of the head.

LECTURE VII.

The Selfish Propensities—Vitativeness, or Love of Life—Misery preferable to Non-Existence—Anecdote—Combativeness—The Steam-Engine of the Mind—Its Uses in all Effort—Abuses—Its Natural Language—Destructiveness—History of its Discovery—Its Legitimate Uses—Necessary to Moral Effort—Abuses: Hatred, Cruelty—Education—Interesting Case of a Boy—Secretiveness—Its Nature and Uses—Indian Shrewdness—Alimentiveness—Function of Alimentiveness—Acquisitiveness—Its Stimulus to Action—Its Labours and Rewards—Mammon-Worship.

THE SELFISH PROPENSITIES.

THERE is a class of faculties which regards *self* in its wants and dangers, and contemplates its protection and their supply; and devoting their energies in the present mode of life chiefly to the physical being, they are called *Selfish Propensities*. The first of these is

VITATIVENESS.

It is the love of life. All know how deeply rooted in our nature is this principle. Man is a dear lover of life. Life to him is sweet. Though it is filled with pain; though overarched with clouds from which leap the live thunderbolts of pain and death; though thorns invest him at every foot-fall; though sin spread wretchedness on every hand, and grief and sorrow mark him as their victim, still he clings to life. It is a blessed boon. He loves it well. It has a thousand dear objects, a vast variety of beautiful things; and though it is canopied with clouds, the rainbow of promise spans them all. It is Vitativeness that gives

this love of life. This is a principle in the mental constitution
as much as the love of offspring, or the love of companions or
friends. The very existence of this affection, as a part of the
spiritual man, is to my mind a proof against the doctrine of the
cessation of being, and in favour of the immortal nature of
mind. It is the life of mind that it loves; the life of the
thinking, enjoying, loving principle.

This affection is manifested through a cerebral organ, as is
every other mental faculty. Some have supposed this to be an
inherent principle of all affection; but we find that in some
individuals it is very strong, while in others it is comparatively
weak. Some will cling to life though bereft of every enjoyment, with a strong and singular tenacity. For nothing will
they give it up. To them there is nothing that looks so awful
as a cessation of being. They would have life continue though
its breath be drawn in misery. The celebrated case of M'Gregor,
recorded in some of our school-readers, illustrates the power of
this affection in some instances. One man once assured me
that endless misery of the most excruciating kind, without
hope of redemption or relief, was to him far more tolerable to
contemplate than absolute destruction, or cessation of being.
To him nothing was so absolutely shocking as nonentity. He
was not a believer in endless sufferings in any form; so we
may conclude that he spoke the full sentiment of his heart. The
strong love of his being was his love of life. Others have told
me that in destruction they could see nothing particularly
terrific. They would prefer to live; but rather than suffer
much, they had rather cease to exist. So that this love is
like all others, strong in some individuals, and weak in
others. Those in whom it is very strong generally cling to
this life with a great deal of tenacity, even though they have
the fullest confidence in an immortal existence. They love
life in any form, and would prefer to exist for ever in this
world, rather than exchange this for one they had every
reason to believe better. Persons in whom this love is very
strong will live under circumstances that would destroy the
life of those in whom it is weak. They love life and deter-

mine to cling to it and retain it. By this determination they
ward off the shafts of death.

I once heard a story of a maiden lady of great wealth, who
was very low with a dangerous disease, and was not expected
to recover. She was considered, and supposed herself, at the
very door of death. She occasionally overheard some of her
anxious relations conversing about her wealth, counting the
several shares, and the amount of each. The good woman's
indignation was stirred, to think that, in the very hour of her
death, her best pretended friends were thinking not of her
loss, but of her *money* which she was leaving to them. In a
moment she determined that they should not have it; and, to
their utter astonishment, ordered away her physicians, and de-
clared that she was not dying, and would not die, but would
live and take care of her own money. The story says that she
recovered rapidly, and lived many years, to the great annoyance
of her friends. There is no doubt but that a strong love of
life does much to prolong this present existence. If one loves
to live, and determines to live, disease must be powerful that
will carry him off.

The organ of Vitativeness is located just behind the ear,
nearly under the mastoid process. When it is large, it extends
back of this process, and gives great width to the head behind
it. Its size and strength can generally be readily determined.
Care should be observed that the mastoid process be not mistaken for this organ.

COMBATIVENESS.

Directly above this is Combativeness, the proper and zealous
defender of life and its rights. Vitativeness loves life, and calls
on Combativeness, its next-door neighbour, to defend it at all
hazards; and so zealous is Combativeness, oftentimes, in its
defence, that it will expose life to the most imminent dangers,
to maintain its own position. The proper office of Combativeness is not to *fight*, but to give spirit, point, ambition, zest, and
fire to the character. Its main object is to act as a spur to the
other faculties; to goad them on to activity, to exertion, to
vigorous efforts, to daring exploits, to bold attempts, to brave

encounters, to great undertakings. It is the active, zealous waker up of the soul. It applies its torch of fire to every faculty, and stimulates each to a flaming life. A mind without Combativeness would be like a steam-engine without fire; a cold and dead association of mighty but sleeping powers. It would be too dull, too lifeless, too insipid for any of the active duties of life,—too nearly dead to be great or good, tender or kind. It could neither love nor hate, think nor act, with any force. It would move at a snail's pace, and be a perfect dump —actually good for nothing. It would scarcely give life enough to the body to cause it to breathe sufficiently to live. While large Combativeness would wake up every energy of the soul, animate the affections with a flame of fire, and the intellect with a torch of light. It would stir the blood to a perpetual fever-heat. It would give fire and zeal to all the noble aspirations, fervency to prayer, brilliancy to hope, tenderness to love, warmth to benevolence, vigour to morality, earnestness to religion, activity to business efforts, and a general vigour and animation to the whole life.

Combativeness can never supply the place of intellect, but it will often whip up a small intellect to great exertions, and cause it to wear the name and badge of greatness; while a mind naturally powerful will lie through life in dormancy and lifelessness, and actually rust out its gigantic powers, for the want of the stirring impulses of full Combativeness. It is often the case, that a small brain acts with great energy, and performs a vast amount of mental labour; while a large one will accomplish but very little, when the reason for this difference is wholly found in the awakening influence of Combativeness. It is the faculty that enkindles the impassioned desire to overcome all resistances, to surmount all obstacles, to get round and over or through all barriers, to conquer all enemies, and to hold a triumph over all victories. It is the faculty that craves, succeeds, and rejoices in it. Hence it is of vast importance in all enterprises. Are you treading the paths of science? Combativeness is necessary to clear the way, to remove obstacles, to break down barriers, to give zeal, and fire, and vigour in the

pursuit of its various objects. Are you prosecuting the claims
of business? It is equally necessary to stir in you the strong
desire for success, to enable you to cope with competition, to
brave opposition, to fear no danger, and to press vigorously
and boldly onward to the attainment of your object. Are you
in the practice of any profession? Its stirring voice you have
need to hear and heed, or feeble will be your professional
efforts, low your professional aims, weak your professional
talents, and small your success. Do you court the rich enjoy-
ments of social life? Combativeness you need, to enable you
to provide a home and its comforts, and realise the exquisite
and refined pleasures it affords to the affectionate heart. Would
you lead a life devoted to goodness, to morality, and religion?
Then the warm fire of this faculty should be enkindled within
you, to stir your soul to vigorous efforts in a divine life, to give
you zeal according to the excellency and the grandeur of your
work, and the fervency of spirit, and strength of aspiration,
which will make permanent the deep and glorious desire for a
God-like life.

The necessity and use of Combativeness, then, are manifest
at once. But it is liable to the greatest possible abuses. There
is danger of its flame rising too high, of its fire becoming too
hot. Then it exhibits itself in a flaming passion; then it pours
forth a volley of angry words, heaps malediction upon malediction, turns its possessor into the image of a tiger, ceases to
become an inspiration for good, and deforms, harasses, and
degrades the whole soul. Its abuse is anger in all its ten
thousand forms — fighting, quarrelling, contending, fretting,
scolding, complaining, fault-finding, vexing, teasing, harassing,
denouncing, ridiculing, abusing, discomforting, &c. In characters where it is strong, it is abused unconsciously. It engenders
a habit of sharp speaking, a pert and tart kind of unpleasant
fault-finding, which is very annoying to others, often planting
a sting in their bosoms which they cannot expel. It gives
the ability and the disposition to carry on the tongue a long,
sharp dirk, something like the dagger which the serpent carries;
and it is run remorselessly into everybody's heart that happens

to do or say anything that does not exactly please. In characters where it is strong, it gives a wonderful disposition and
ability to use sharp, sarcastic, venom-toothed words ; words
that bite, and sting, and corrode ; caustic words, that eat into
the very quick, and make one's soul smart as though an adder
had stung it. This faculty is very generally abused in giving
frequent utterance to such corroding words and sentences, and
giving birth to the feeling out of which they grow. The natural
language of this faculty is very plain. It gives a quick, side-
way snap to the head, with a little cant backward. It gives a
hasty, quick tread, and a kind of snap to all the motions of the
body ; a quick, hasty, clipping manner of speaking ; a darting,
vivid expression to the countenance, and a restless and impa-
tient manner to the whole person.

DESTRUCTIVENESS.

The nearest kindred to Combativeness is a faculty named
Destructiveness. By common consent, among phrenologists,
it is called by this cognomen ; but it is most evidently misnamed·
It is named in view of its *abuse*, rather than its *use*. The great
leader in phrenological science was long doubtful whether he
should call this faculty Destructiveness or not. He disliked to
believe that man was a destructive being. Yet he saw him, in
his wars, murdering his own species ; he saw him, in his sports,
destroying the animals about him, often doing it for the merest
pastime ; he saw him, for aliment, devouring the very animals
which did him the best service. And what could all this mean ?
He must have a faculty in his mind which delights in destruction.
Often, in children, we see a species of delight in destroying
their toys and playthings. These considerations induced Dr.
Gall to designate this faculty *Destructiveness.*

But it appears to me that it is named with reference to its
abuse. It is true that man does all these things ; but he does
them in *abuse* of his nature, in abuse of the very faculty which
enables him to do them. Similar remarks might have been made
on Combativeness. That faculty is also named from its abuse.
Man is a combative, fighting being, under the abuse of his

nature. The true and legitimate office of these faculties is to give energy to the character ; to give force to the action of the other faculties. Combativeness is the fire, while Destructiveness is the steam of the spiritual engine. Combativeness kindles the fire which raises the steam, by which the whole mental apparatus is forced into powerful and active labour. And a mind without Destructiveness would be just as useless as an engine without steam. Force of character, energy of spirit, power of action, are conferred by this faculty. It is the spring-source of that prime and cardinal virtue, *Perseverance.* We have often heard the praises of perseverance ; we have heard of its energy, its labours, its tireless arm, and unflinching zeal. These are but so many praises of Destructiveness. Perseverance, however, it should be remarked, is a compound virtue, formed by Destructiveness and Firmness ; but its active element, its element of power, comes from the faculty of which we are treating. All men of energy, of bold and resolute determination, of vigorous action, of strenuous endeavour, of thorough-going force, are strongly endowed with this faculty. It gives power to the will, vigour to thought, and success to action. It is pre-eminently the faculty of *success.* It digs, forces success out of every enterprise it undertakes. Look around you at the successful men in the conflict of life ; they have strong and active Destructiveness. It is necessary in every business and pursuit, even in the pursuit of moral good. The moralist must have it strong, or his morality will be weak and sickly. The religionist has it strong, or his religion will be but a faint desire. It will never show itself in noble actions, in self-denial, in strenuous spirit-struggles for good. In no work of life is it more absolutely necessary than in the self-sacrifice and discipline imposed upon man as the noblest and last duty of religion. To overcome the undue exercise of the selfish sentiments and propensities, to curb the appetites, to bridle the lusts, to resist temptations, and to labour with a manly boldness and vigour for the high vantage-ground proposed by religion, is a work of indomitable energy. The reformer has great need of this power of mind. He has to oppose old errors, old practices, time-honoured usages, and

H

work his way against the strong tide of popular sentiment, and
the mighty barrier of popular prejudice. Silent will be his tongue,
and palsied his hand, if he is not strongly endowed with the
energy and power of Destructiveness.

Every man, every woman, has need of the strong impulse
given by this stirring, pushing, daring, restless energy of soul.
It is the origin of efficiency and thoroughness of spirit, and
the sworn enemy of tameness.   It nerves the arm with power,
sharpens the intellect, stimulates the moral sentiments, fires the
affections, presses into action every power of the soul. It gives
not a fitful flame of energy, but a steady burning impulse.
Most essential is this faculty of the human mind ; yet it is
liable to great abuses.   It is a strong impulse ; a powerful passion ; and when not held by the strong rein of self-restraint, it
often overruns all bounds of moderation, and bursts out in violent passion, in deep anger, in boiling resentment.   When it is
stirred to hatred, it is deep and uncontrollable.   It is the mad-
ness of the bull-dog, the deep, vindictive rage of revenge.   It
is the feeling that holds grudges, that cherishes resentment,
that burns in a fire of perpetual hatred.   It never likes to bury
the hatchet.   It wars against forgiveness.   It is the seat of
everything that is black and revengeful in malicious hatred.
The organ of this powerful faculty is in the base of the brain,
just above the ear.   It is in the centre of the basilar brain.
When it is large, it gives great width to the head ; and when
very large, it makes the head nearly round, like the head of the
bull-dog.   Beware of a large round head ; it has a gulf of tartarian flames within it.   The head should not be too flat nor
too round.   If too flat, it will lack energy ; if too round, it
will be very liable to run into the extravagant abuses of Destructiveness.   Great care should be taken in the cultivation of
this faculty.   It should be trained with tender solicitude, and
made to work for the higher and nobler sentiments.   When it
is very large in children, the only way of proper treatment is
to mould their spirits with kindness.   I once tried an experiment on a little boy about four years old.   He was a sweet,
active, and usually a good child.   It was in the early part of

my teaching, when I had not so much control over children as I afterward obtained. I observed, on the first day that this boy entered school, that he had enormous Destructiveness, and I feared that I might have trouble with him. Several weeks passed, and he proved himself as pleasant a scholar as I ever had. Now, Phrenology, thought I, here is your test? Notwithstanding the child's perpetual pleasantness, I doubted not the testimony of this science, that cannot lie. At length the day of trial came. The little fellow wished one day for a privilege that I thought it not proper to grant him, and in refusing to comply with his request, I accidentally did it in such a way as to offend him. He commenced a loud and boisterous cry. I coaxed, pleaded, entreated ; tried every means within my power to pacify him, but all to no effect. I tried letting him alone, but that made him all the worse. Again and again I tried to flatter and coax him to quietness, or to turn his attention to something else ; but no, he grew every minute the more wilful and outrageous. After some half-hour's fruitless effort in the way of kindness, I thought I would try the force of threats. These succeeded no better. I then resorted to the rod ; but this only added fuel to the flame. I whipped him till I was afraid I should do his body a positive injury if I persisted. I then imprisoned him under a desk till he cried himself to sleep. He slept some half-an-hour. He then awakened in as much of a rage as when he went to sleep, and commenced his bellowing cry. I went to him and endeavoured to pacify him, but all to no effect. After crying at the top of his voice for some half-hour longer, I went again and spoke in the sweetest tone I could use, and lo ! it was like a miracle. He looked up and smiled, as does the teary sky after a thunder-storm. He wiped his face, took his seat, and looked as happy as one first delivered from a dreary prison. His violent passion lasted nearly two hours. Had he been strong enough, he would have worked his way to the desired object, had it lain through a stream of blood. This illustrates the abuse of Destructiveness, when very large.

SECRETIVENESS.

Between the organs of Destructiveness and Cautiousness lies the organ of Secretiveness. Man is in great need of a faculty which shall enable him to conceal his feelings, to hide them from the public gaze. If every feeling of his heart, every thought of his intellect, and every suggestion of his propensities were acted out, and the whole inward man, in all its various states and changes were exhibited in the outward life, what a strange, ludicrous life he would exhibit! Who would have the world know the secret whisperings of his propensities? the contentions and struggles that go on within him? Then, how could man form his plans of life, do his business, control his affairs, if the suggestion of his every faculty was carried at once into the outward life? The truth is clear, that a concealing faculty is absolutely needed. It is necessary for him to hang a curtain around his soul, and do his planning behind it. Secretiveness affords this curtain. When very strong, it is the seat of hypocrisy, lying, cheating, deceiving, trickery, stratagem, and all kindred vices. It gives a low shrewdness, cunning, and deceitful sharpness. It is the leading power in the Indian character. It is the source of Indian shrewdness and cunning, watchfulness and deceit. When combined with large Acquisitiveness and small Conscientiousness, it makes a thief. You will see its best illustration in the cat tribe of animals. See how cunning and shrewdly they lie in wait for their prey. When this organ is large, combined with a large development of the other side organs, it gives a sharp business tact to the mind, a planning, scheming, contriving disposition in all business matters, and almost always makes a successful character. Its natural language is sly circumspection, watchfulness, a cattish expression and action; a still, careful walk; a low, sly tone of voice, frequently falling into a whisper; a disposition to whisper in the ear, to step aside to hear the most trivial thing, &c. It is a very useful faculty, but is liable to abuses. Great care should be taken in guarding it well. It needs a world of good training.

ALIMENTIVENESS.

No fact is more clearly written in the history of the human species, and in the experience of man, than that man is an *eating* and *drinking* being. Eating and drinking is the business of life.· In every season of life he eats. In every age of the world he has eaten. In all countries he eats. The high and low, the rich and the poor, eat. The king and the beggar, the *belle*, the washerwoman, the professional man and the chimney-sweep, all eat. Surely we need no more proof than we have, that man is an *eating being*. He devours every green and every living thing. Who doubts that man has a natural disposition to eat? This is a mental, not a bodily disposition. It is the mind that calls for food when the body is in need of it. The body is the mind's subject. The mind must take care of it; preserve it; guard it; supply its wants. It is continually subject to the wear and tear of life. This continual decay is supplied by food. The organ which gives appetite is in the base of the brain, and is located just in front of the external opening of the ear, and above the cheek-bone. When it is very large it gives a full, swelling appearance to the sides of the head, in front of the ear; a widening from the eyes back. It makes a good eater, a lover of good victuals. It gives a warm respect for the table, and especially for its precious burden. Feastings and fast-days are to this faculty great and memorable occasions. In woman it gives an excellent ability for the culinary profession, and makes her an accomplished mistress of the kitchen, cooking-stove, and dining-room. If time would permit, I should like to speak at length of the *abuse* of this faculty. Men, instead of eating and drinking to sustain life, are eating and drinking to destroy it. They are eating themselves into the epicure's grave. They gratify appetite at the hazard of health, peace, wisdom, morality, religion, spiritual progress, happiness, and everything else that is good.

ACQUISITIVENESS.

That man has an acquisitive disposition, no one denies. It is proved by his life, by his large estates, by his daily objects

and labours, by his service for Mammon. Acquisitiveness is
the property-loving instinct. How deeply seated is the love of
gain in our natures! and how much does it influence us in our
daily avocations! Look out upon the world, and see the scramble
for wealth. In every continent, and on every island where
civilised man is found; in every country, nation, state, town,
neighbourhood, men are eager for gain. Days, and months,
and years are spent in earnest efforts for acquiring wealth, in
almost every one's life. Look at Commerce, with her million
sails, her thousand engines, and her innumerable flotillas. Behold Agriculture, with her harvest-fields and granaries spread
all over the world. See Mechanism, plying her myriad spindles,
hammers, saws, and industrial engines. What is the mightiest
spring of action in all this busy world of labour? Is it any
natural want or necessity? No; for the wild man of the forest
can more easily supply his natural wants. It is the love of
gain.

The love of distinction, no doubt, has much to do with it;
but the strongest moving power is the faculty of Acquisitiveness. Open a gold mine, and it sets the world crazy! Discover
a new article of commerce that offers opportunities for great
gain, and what a pulse will beat throughout the civilised world!
Open a new territory, and the tide of emigration sets toward
it as though an irresistible attraction were there. Anything
that offers a fair chance of great gain will upheave the very
foundations of society. Such is the power of the Acquisitive
faculty. It cries "Get, get, get!" It is as trong and powerful
impulse, stirring the whole soul often times to the most vigorous and determined action.

This faculty is absolutely necessary to man's well-being and
happiness. He has need of an Acquisitive ability to enable
him to lay by a sufficient store of life's necessaries to guard
him against want, to support him in the hour of sickness, misfortune, and age; and to give him an opportunity to bless the
needy, feed the hungry, clothe the naked, instruct the ignorant,
and do all acts of goodness which his heart shall dictate. It is
necessary for man's well-being that governments be founded,

laws enacted and executed, schools established, asylums made, books and papers published, roads built; institutions, churches, hospitals, &c., erected—all of which would remain undone, were it not for the Acquisitive faculty. The means by which all of these noble objects are accomplished are garnered up by Acquisitiveness. Then, again, Acquisitiveness, like all the other faculties, has its twofold office, its material and spiritual use. There is truth to be garnered, knowledge to be hoarded, wisdom to be gained, character to be acquired, and all the inestimable treasures of the heart to be secured. The intellect is to be stored, the moral man is to be enriched with the endowments of righteousness; and he must have its diamond wealth. All these things are to be acquired. Acquisitiveness affords a strong stimulus for their attainment. This is the ultimate end and object of the Acquisitive faculty. This is its spirital office. It is a grand, noble faculty designed to work out for man a glorious end. It should be properly cultivated and directed to its spiritual objects. It should be made to work for the acquisition of whatsoever is good, beautiful and true. It should not be permitted to waste its energies on earthly objects—merely on gold, wealth, perishable things; but taught to look beyond, to the permament and enduring wealth of mind.

This being a strong and active faculty, it is very liable to abuse. Its chief abuse is "Mammon-worship." It loves "the dimes," and is always ready to fall down and worship a golden calf, or any image that is made of gold, or looks like gold, or can be converted into gold, or anything that gold will buy. It causes penuriousness, littleness, meanness, tightness, or tight-fistedness, and all kindred vices. It makes the miser lean, gaunt, niggardly as he is—a moneyed lunatic—being subject to a morbid action of the organ of this faculty.

When it is large, and combined with small Conscientiousness, it causes theft, robbery, and murder for money, and all kindred crimes. It has a wonderful power of magnifying money. It will make a shilling look as large as a full moon, and a five-shilling piece like a moon through a thousand-powered telescope.

"Take care of the pence, and the pounds will take care of

themselves,"."Time is money," "Save as you go," are adages that
had their origin in this faculty.   When it is large, its possessor
is a money-maker, a money-saver, and when it is small, he
possesses but little ability to keep what he earns : money will
slip through his fingers.   He will find a thousand ways to spend
it.   Give him the best opportunities in the world, and he will
not get rich.   Let him earn ten thousand a year, and it will not
supply his wants.   Many such persons we find.

The organ of this faculty is located just above and a little
back of Alimentiveness.   It is directly in front of Secretiveness.
When large, it gives great width to the head, just above and in
front of the ears.

LECTURE VIII.

The Perfective Group of Faculties—The Faculties of Civilisation—The
Manufacturing Talent—The Mechanical Organisation—Tune, its Location—The Power of Music—The great Masters in Music—Music the Language of the Soul—Ideality, Perfection, Imagination—Beauty the Basis
of its Inspiration—Ideality the Poetical Faculty—The Foundation of
correct Taste—Sublimity—Objects which inspire Sublimity—Its Influence
on Poetry and Oratory—Imitation—Its Function and Location.

SEMI-INTELLECTUAL SENTIMENTS.

WE approach now a more delightful region of mind.   We enter
upon enchanted ground.   The charmed region of the imagination is before us.   Here is the mysterious, wonder-working
wand of mechanism.   Here is the thrilling, enrapturing power
of music.   Here is the charm-bearing sense of the beautiful,
and the powerful sentiment of sublimity.   This is called the
semi-intellectual region of mind.   When strong, it gives a wonderful power of appreciating whatever is beautiful, complete,
finished, exquisite, grand, lofty, and majestic.   It is the poetic
corner of the soul.   It gives all those fine flights of fancy, those
lofty careerings of the imagination, those exquisite revealings
of intuition, those sun-flashes of poetic genius, which sparkle
along the pages of the masters of the greatest of human arts.

Poetry! thou art the offspring of genius, the human child with angel powers, the earthly type of heavenly things; and thou art born and cradled, nursed and grown in this department of the mind. Here, close by the side of the throne of reason, under the very droppings of the sanctuary of religion, just above the flames, but not above the subdued heat of the passions, thou hast grown to possess thy matchless charms, thy divine beauty, thy spirit-stirring power. Out from this focus of awakening energies, this meeting-place of heaven and earth, thou hast come, wrapt about with thy charm, shining with the light of thy soul, to ravish and bewilder the senses, and the souls of human kind. Come, worshipper of the mind's great Author, witness here the infinite beauty of His arrangements! See what a star-girded palace He has prepared for the birth-place of poetry! Around it stand the watchful sentinels of reason, charity, religion, and righteousness. Below it dwell the tide of earthly passions. Thus, poetry is evidently heaven's evangel to earth—God's minister to men. And this department of mind is of inestimable value in elevating, refining, redeeming human creatures. It wraps earth in the charm of heaven, robes the material in the garments of the spiritual, and gives man a foretaste of the life to come. Excellent and holy is the office of the semi-intellectual powers. They are the servants of the higher, and minister to, guide, and sanctify the lower faculties. But we must approach them singly, and therefore first call your attention to that great agent of civilisation, Constructiveness.

CONSTRUCTIVENESS.

The first and the lowest of these is Constructiveness, the making, building, constructing, mechanical ability, which civilised man exhibits in such multiform splendour. It is useless to particularise in this mechanical age. The wonders of the Crystal Palace of London, in which the world's great industrial fair was exhibited, speaks for this faculty. The houses in which we dwell, the costumes we wear, the carriages in which we travel, the towns and cities in which we congregate, the farms we cultivate, the millions of labour-saving machines with which

we ply the busy concerns of life, the shops that dot our continents all over, and the ships that checker our seas, are all but
so many living voices of this faculty. Man is said to have been
made in the image of his God. God is a mechanic of infinite
skill. He built the universe, an infinitely sublime spectacle of
mechanical design and execution. Man builds houses and con-
structs implements of labour ; and how sublime a spectacle does
he exhibit in his mechanical operations. What would man be
without Constructiveness? A wild man of the woods at best.
This faculty has administered as much to man's elevation,
perhaps, as any other in his mind. It is the first that is called
into requisition in the act of civilisation. It is that which
furnishes us with all the conveniences, and nearly all the
comforts, of life. It is Constructiveness that makes the mechanic, and the mechanic almost makes us. One moment's
consideration will show us how much we are dependent upon
the mechanic. All that adorns and embellishes life, we get at
the hands of the mechanic. We can scarcely think of anything
but air, water, and wild fruits, for which we are not indebted
to the mechanic. Useful and honourable then is the true
mechanic.

When the organ of Constructiveness is combined with large
Perceptives, and a keen and active temperament, it constitutes
a genius in mechanism. This combination usually forms an
artist, or a workman at some refined species of mechanism.
All the prominent artists that I have seen, have this combination. Thus, if God makes a man an artist, He writes out that
fact on his head and in his nervous excitability. Go into a
school of young men and pick out the artists, the genuine mechanics. You can tell them as well as you can tell a white
man among a crowd of black ones. They have prominent Perceptives, wide heads in front and above the ears, or about the
temples, with active temperaments. With these the principles
of mechanism are a perfect joy. They behold an exquisite piece
of workmanship with ravishing delight. A machine-shop, a
factory a workshop, where finished mechanics or artists are
plying their busy trades, is a place of the richest entertain-

ment. They delight to work with their own hands in making and constructing such things as please their tastes. And they have ability to work. They can work, as it were, by a sort of intuition, work as though the genius of mechanism directed their hands. Everything that comes from under their hands has the finish of perfection upon it. They know how to make everything just right. They *know* how, they don't have to *learn* how; while those who are the very opposite in form of head and in constitution can make nothing. I knew a man of this latter description, who, his wife said, could not make a button to hold a door—could not make anything; but fortunately for him and his family, he was a great man at digging potatoes. It was said that he could dig and put into a cart seventy bushels in a day. As wisdom will have it, every man is made for something.

The organ of Constructiveness is directly in front of Acquisitiveness. Its size may generally be told by a glance of the eye from a front view of the sides of the head. When it is large, it gives a swelling appearance to the sides of the front head.

TUNE.

Just above and a little forward from Constructiveness is located the organ of Tune, which manifests the faculty which delights in musical sounds, in concords, in harmonious strains, in sweet, floating melody. The charm which this power of mind feels in the presence of sweet, concordant sounds is perfectly transporting and intoxicating. It carries its possessor into the third heaven of bewildering joy. When united with large Time —the organ of which is located just in front of Tune, the office of which is to recognise the passage of time, and count its moments as they roll, and which is ever pleased, yea, delighted with the metrical flow of numbers—its gratification affords one of the most intoxicating feasts of delight of which mortals are permitted to taste. It is then that the rich and exhilarating charm of music is felt, and its powerful spell is thrown around the soul; then that it rules with the tyrant energy of a master-passion the whole inner court, and gives the mind up a subject to a fever-fit of delight. The power of music we nearly all of us understand; it is almost universally felt. How it works within us a wild delirium of joy, thrills the trembling nerves of feeling, pours along the current of life its floods of soul-stirring harmony, and breathes an influence, at once wild and sweet, through the whole being.

There is no doubt that this power is conferred by a distinct faculty or faculties of mind. In some it is very strong; in others it is very weak. The power to make or appreciate music is a compound power, for music is a compound made up of harmonious *sounds*, uttered in flowing measures of *time*. To make it, or appreciate it, two faculties of mind are required; one to give utterance to the proper sounds, the other to measure those sounds into the correct time, or into their proper divisions or feet. In persons in whom both of these faculties are strong, music is an intuition; a rich, natural spontaneity; especially if the temperament is strongly mental, so as to give that exquisite delicacy of taste, perception, and feeling, requisite to detect and make all the nice varieties of sound and harmony, which constitute the supreme charm of music.

Blest are they who have these organs well developed and have properly cultivated them. They have within themselves a source of thrilling and elevated pleasure, and one the tendency of which is to refine and bless their souls. But these faculties like all others, are the subjects of cultivation. In order that their best influence may be exerted, their highest pleasure realised, and their grandest powers manifested, they must be subjected to a long, rigid, and tyrannical discipline. Look at the masters in musical science, Haydn, Olè Bull, Jenny Lind, Catherine Hayes. They are what they are by their untiring diligence in cultivating their powers. They determined to, and did persevere against all obstacles. The best female singer I ever heard, told me that she had spent the best part of eight years in the study of music. To be great in this science requires the same exertions that it does to be great in any other department of human exertion. Who are our eminent lawyers, physicians, statesmen, divines? Those who have devoted a half century of untiring effort, study, and practice to their several professions. The men of labour, of toil, of unfaltering perseverance for the attainment of their objects, are the men of success. The musical talent is, perhaps, as nearly universal as any other, even as the talent for the use of language, and yet but few, comparatively few, are musicians. All men make a free, and nearly all a copious use of language ; all men talk, express easily their thoughts. Why then should not all men sing, be musicians, appreciate and enjoy music, make melody in their hearts? There, of course, must be a great diversity of musical ability, as there is in every other natural talent. But this is no reason why our musical natures should not be cultivated. It is rather a reason why they should be cultivated, that all may be more equally developed and enjoyed. Every heart should be the home of music. Every home should be an orchestra for sacred, and an opera for sprightly, gladsome music. The richest sentiments of the heart are to be sung. The holiest feelings of the soul find their best utterance in music. The warmest, purest affections express their fervour in song. Music is the natural language of a full soul. The mourner, yes, even the mourner, loves the mellow, grief-laden cadences of slow and solemn music. The patriot expresses his bursting joy in the rich notes of the bugle ; the warrior shouts in answer to the stirring fife and drum ; the worshipper rises in praise with the swelling strains of the organ ; the lover melts to the soft tones of his ravishing lute, whose very melody seems the sweet breath of love. It is proper, then, that all should be master of this glorious science. One day which is to come, all must be taught, and will willingly, yea, gladly, join in the anthem of redemption. All souls will one day be full of love, and praise, of heavenly aspiration, and glorious sentiment. At once, then, they should enter upon a preparation for that great day of the fulness of glory ; at once they should begin to taste a prelibation of that flood of angel-music which is to be poured into the ear of creation's King. While the infant mind is untarnished with sin and corroding care, this heavenly element should be awakened to a chorus of harmony that shall ever swell above the harsh discord of life. Early should the soul's great powers be developed for the work of the immortal world, and the enjoyment of immortal felicities.

IDEALITY.

Just above Constructiveness lies the organ of Ideality. It manifests the sense of beauty. It is the central power of what the old metaphysicians called imagination. Its office is to confer the idea of beauty, perfection, completeness, finish. It is joined closely with Constructiveness, and assists it much in inspiring its conceptions of perfectness in mechanism, both in principle and in action. Every mechanic should have well developed Ideality, in order that he may be a finished workman, may give completeness to his work, correctness to his taste, and harmony to the products of his skill. Ideality has a wide range for its operations. " No pent up Utica contracts its powers." The whole boundless universe is the scene of its delights and labours. From the starry sky to the flowery earth; from the radiant pole to the flashing meridian; from the tiny insect to the omnipotent God, it loves, admires, and glories in the beautiful.

The completeness, finish, harmony, and perfection of all the works of nature are the food on which it delights to feast. The flower is beautiful, and Ideality loves it. The plant is beautiful, and Ideality tends, waters, and cherishes it, to preserve and perfect its beauty. The landscape is beautiful, and Ideality gazes upon it with a kind of delirious joy. The rippling brook, the leaping cascade, the silvery stream, the sloping hill, the flowery lawn, the leafy grove, the round-topped tree, the graceful willow, and the clinging vine; the morning dawn, the white, floating cloud, the golden sunset, the gay birds of plumage, the radiant bow of promise, the starry canopy of night, are all beautiful, and they are all the dear, darling children of Ideality. It loves them as the mother does her babe; and it never looks at them without feeling a strange, wild delight, such as the lover feels when he gazes into the eyes of his beloved. All beauty works an inspiration in this faculty. It loves the beautiful in art as well as nature. Beautiful buildings, fences, yards, pieces of mechanism, such as furniture, carriages, dresses, books, statues, pictures, paintings, shells, &c., awaken its delight and charm it into a passion of pleasure. It is the faculty which makes people well-nigh crazy in their admiration, wild, uncontrollable, which bursts out in exclamations, ejaculations, and delirious expressions of delight which nothing can restrain, and which, without respect to time, or place, or occasion, or company, will give utterance to the thrilling and beautiful emotions that rise within it. When it is strong, with a very refined temperament, it gives richness, deliciousness, exquisiteness, delicacy, and refinement to all the feelings, which completely elude the touches of the most delicate and powerful pen in their expression. They are to be *felt* by souls so fortunately endowed, but are never to be described. There is a beauty in this faculty, in its delicate and exquisite sense of perfection, which far transcends all material beauty, and which the faculty itself greatly admires and loves in others. This faculty and temperament is the spring-source of all true poetry. It loves, writes, feels, thinks poetry. It is always poetical, and though it writes prose it will breathe into it the spirit of poetry. Its conceptions will always be fine, chaste, and pure, but strongly tinctured with what cooler and coarser minds call extravagance. Every writer that has this power in great strength will show it in his writings. He will be flowery, florid, extravagant, generally verbose, full of hyperbole and figures. Everything must be finished to his taste; every sentence completely rounded, every word just the one, every epithet the strongest, and in the superlative degree. This power, properly cultivated, gives the highest charm of perfection to one's style. It loves *style* oftentimes more than *sense,* and if not well disciplined will frequently sacrifice the latter to the former. It generally gives quickness, sprightliness, gorgeousness to one's style, which makes it generally pleasing. Young writers should guard against its abuse.

When it is united with powerful intellect, it gives beauty and richness to intellectual pursuits; adds a new fire to the flame of intellect, and generally quickens it to wider, vaster, and more comprehensive views, and tinctures its philosophies with the theoretical, the visionary, the ideal, and extravagant.

When united with strong propensities, it fixes its attention chiefly upon objects of material and physical beauty; beauty of person, form, colour, activity, motion, energy, &c. When it is united with strong moral sentiments, it admires, most of all things, the beauty of character, virtue, purity, charity, honesty, righteousness, religion; the beauty of spiritual and immortal things. This is evidently the highest use and end of Ideality. When united with strong affections, it admires warmth, ardour, and strength of friendship and love. Thus the character that Ideality assumes in any mind will always be greatly affected by its combinations; but it will always love something finished, complete, perfect in its kind. It dislikes the coarse, the low, vulgar, inharmonious, disjointed. It is pained beyond expression with all such things. Ideality has much to do in the formation of correct tastes. And in the gratification of its tastes it is very particular. It must be satisfied, or it is greatly pained. Few can imagine the pain, the actual anguish, that coarse and vulgar expressions of taste give to refined Ideality. So it is a truth, that he who has such an organisation is capable of the most exquisite happiness; but is also capable of the most intense misery. I have observed that Ideality is almost always strong in reformers. It is pained at the misery and wretchedness, vulgarity and depravity, which it everywhere beholds; and is delighted with the views of peace, prosperity, harmony, and excellency which it beholds in its visions and dreams of a perfect world and a perfect society; and it urges the whole mind to co-operate with it in affecting in reality what it beholds in vision.

Perfection is its grand end. For perfection it was made. In perfection it finds its delight. Perfection is its dream of joy, blessedness, and hope. This love of perfection is a part of the immortal mind; and that mind must for ever remain miserable if it cannot reach perfection itself, and behold perfection throughout the realm of mind. This faculty clearly

points forward, in my mind, to a day of perfection yet to be
attained by itself and all sentient and moral beings. Its grand
object in the mind, no doubt, is to secure this glorious end.
It desires perfection in itself, and is miserable at the sight of
imperfection in others. No human being can ever be perfectly
happy while imperfection is found or known to exist in the
universe of God. Glorious are the inspiring prospects given
by the prophetic views of this faculty. It is the prophet of
universal perfection, the inspirer of a strong desire for a perfect
life, a perfect mind, a perfect wisdom, and a perfect love. Let
it be cherished with affection, and cultivated with care.

SUBLIMITY.

Stepping back from Ideality, just at the corners of the middle
head, we meet with the organ of Sublimity, the office of which
is to manifest that power of mind which recognizes the grand,
the vast, the magnificent, and sublime in nature, art, feeling,
thought, and action. The mountain is grand, and sublimity loves
it. The wide, rolling river is grand, and its swelling grandeur
touches this faculty. The thundering cataract moves with
power upon this same faculty. The great ocean, the arch of
heaven, the circuit of the world, the lightning's flash, the roar
of thunder, the tread of armies, the movements of nations, the
great moral movements of God, in creating and controlling
mankind, the work of redemption, the prophecies of a time of
universal concord and peace, of the gathering of all nations,
and the people of countless ages into one home, one spirit, one
wisdom, and one law, are all objects of grandeur and magnificence, and they appeal to the faculty of Sublimity with a
tremendous power. It contemplates these with a lofty delight.
The charm which it inspires is similar to that inspired by
Ideality, save that it is awakened by different objects. It gives
comprehensiveness, vastness to the mental conceptions, plans,
and operations, and inspires strong desires for the accomplishment of something in accordance with its own vast spirit.
When united with strong intellect, it gives width, comprehensiveness, and grandeur to the intellectual conceptions, and gives

a peculiar pleasure in contemplating and studying the sublime works of the universe and its God.

The highest exercise of this faculty, is, doubtless, the contemplation of infinity as applied to God, to illimitable space, and eternity of duration. No other mental power is capable of grasping these ideas, and no one but this contemplates them with pleasure.

When united with strong moral powers, it inspires moral grandeur, sublime views of God, His creatures, and His moral government, and contemplates the grandest possible results to the great work which God is performing in His moral creation. It is greatly inspired by hope, which lives next to it, and hence delights in contemplating the future filled with the grand results which it delights to behold. This faculty, together with Ideality, points forward to a time when it shall find a full and complete blessedness in realising that moral grandeur which the Revelator contemplated, when in vision he heard, "Every creature which is in heaven, on the earth, under the earth, and in the sea, and all that are in them, heard I saying, Blessing, and honour, and glory, and power be unto Him that sitteth upon the throne, and to the Lamb for ever and ever." Sublimity shows itself in an author's style. It speaks its own appropriate language, and gives life, loftiness, and nobleness to the style. Those poets who write under the inspiration of this faculty are the poets who live and will live through all generations. Homer, Milton, Pollock, Bailey, are of this character. Their style is full of strength, majesty, power, and lofty magnificence. Their periods are like the rolling of ages ; their thoughts like glittering worlds ; their rises and falls like ocean tides. They carry us along in their majestic strength, like the storm-god whirling the obedient clouds. There is something more than captivating in such authors. They are absolutely overpowering. They wield the sceptre of an almost divine authority, and toss to and fro the little minds of their readers like leaves in a whirlwind. When this organ is large, it gives great width to the middle of the top head. It is directly in front of Cautiousness. It exercises an ennobling influence in the character, and when strong, makes its possessor despise little, mean acts and thoughts. It wants everything done on a great scale, worthy of a true man.

IMITATION.

Above and a little in front of Ideality is the organ of Imitation. It may truly be called the parrot of the human soul. It seems to be entirely destitute of originality. It is the spring of no new thoughts. It wishes to walk in no untrodden paths. It always wants a guide, a pattern. Without a pattern it is powerless. As a mechanic, it works after a pattern; does what it has seen done, and nothing more. This mental faculty assists the workman very much, for it enables him to imitate everything that he has seen in mechanics,. and this, added to the inventive power, gives a two-fold advantage to the mechanic. A dressmaker having this organ large can make anything she has seen in the form of a dress. The same is true of every kind of artisan.

When this is very strong in any mind, it gives a good talent, and a desire equally strong to imitate everything curious, odd, strange, or new. In children it operates very strongly, and forces them to attempt to do almost everything that they see older people do. Their ten thousand pranks are done in imitation of what they have seen; hence the propriety of a correct example. All the moral force of example appeals to this faculty. We are all of us imitators, and greatly under the influence of the examples which are daily set about us. From the infant to the grandsire are imitators; but the king of imitators is the true, successful actor. All who have been noted stage actors, players, dramatists, have been indebted to this faculty for their eminent success. It is located on each side of Benevolence, and gives a round, wide, full head from Ideality over to Ideality. Let the class examine the heads of good imitators, mimics, &c., and they will soon learn the appearance of the organ when large or small. Parrots, mocking-birds, monkeys, &c., have it large. There are many important lessons connected with this faculty which we must pass over for the present.

LECTURE IX.

The Intellectual Faculties—High Office of the Intellect—Individuality; its Office—Form; its Office—Size; its Office—Weight; its Uses—Colour; its Office and Value—Order, a Law of Nature—The Importance of Order —Its Practical Use—Calculation, or Number—Idiots with Large Calculation—Locality, or Sense of Direction—The Exploring, Navigating Faculty—Science based on the Perceptives—Eventuality, the Mental Storehouse—The Historian of the Mind—Causality—The Question-asking Faculty—Comparison; its Office and Utility—Mirthfulness—Language—Language, the Voice of the Mind—Cultivation of Language.

INTELLECTUAL FACULTIES.

In our excursions through the departments of mind, we come now to the home of the intellectual faculties. We visit the dome of thought, the birth-place of those kingly powers whose majesty is known through earth and heaven, who rule over matter and spirit. They dwell in the frontal region of the head. They occupy the foreground of the soul, indicating their office, which is that of supreme direction. They hold the sceptre of authority. They wear the crown of wisdom. They are the soul's grand leaders up the highway of knowledge. They pave the road to the fields of eternal wisdom. They gather the diamond treasures of thought. They weave for man his crown of light. They unveil to his gaze the beautiful universe, which is the body of the great thought of the Eternal. They uncover the hidden springs of power and motion, which lie concealed in the elements, and objects, and worlds about us. They rend away the material covering which envelops the unseen forms about us, and show us the living, enduring, spiritual reality in which reside all power and true majesty. They read us the history of our grey old earth as they find it written on the everlasting "tables of stone." They explore for us the wild continents, islands, and oceans that form the face of our globe. They open for our study the three grand kingdoms of nature—the mineral, the vegetable, and the animal. They analyze and classify the myraid and multiform objects of each, and show us their intrinsic and everlasting utility and beauty. They read us the great statute-book of nature, its laws, general and particular. They give us all true knowledge of the earth and its objects, the sky and its glory. They do our business, make our laws, establish our government, direct our labour, control our agriculture, commerce, and manufactures ; make our books, teach our children, inform our people, read our lectures, pronounce our orations, preach our sermons, lead us through life, and open for us the gates of glory. Magnificent are their labours, splendid their achievements. They have built the pyramids of thought that stand over our world like colossal piles of light. They have formed the sciences which are the great substratum of human excellence and progress. They have civilised man, and promise to redeem him. They should be man's guide ; they should direct his steps, control his actions, lead him ever ; acting always in obedience to the dictate from the throne of the moral sentiments.

INDIVIDUALITY.

The first intellectual ability necessary to man, as a sentient being, is the ability to recognize things and facts. Nothing can be made without materials. No conclusions can be formed without facts ; no reasoning done without premises. Man has ever loaded himself with eulogies for the possession and exercise of the power of reason, and devout men have deeply thanked God for the gift of this wonderful ability. But all have not perceived that this power is dependent upon another which must antecede this in all inquiries. That antecedent power is the searcher of facts, and is called, in phrenological science, *Individuality*, because it recognizes individual things and existences. Its office is that of fact-gatherer. It is the seeing, perceiving faculty. It takes cognizance of things as things, of facts as facts, without relation to any other things or facts. The properties and qualities of things it does not recognize, but simply sees everything as an individual existence. Nor does it name or classify the objects which it sees. Those works are left to other faculties. In the old systems of mental philosohpy this faculty was called " simple perception," and very properly named it was. This is the grand inlet to the intellect. It is both the telescope and microscope of the soul. It is surrounded with other faculties which determine the qualities and properties of objects. Its organ is located in the centre of the forehead, just above the eyes. The eyes are its servants. It uses them for its work of observation.

FORM.

Below Individuality, at the root of the nose, is located Form, the office of which is to observe the *form* of the objects which Individuality sees. It sees and remembers all the peculiarities of figure, shape, and feature of whatever passes before it. Hence it observes accurately the form of faces and persons, and remembers them from their forms; the form of houses, horses, towns, &c. Those who have this organ large usually have an excellent memory of persons—seldom forget any whom they have once seen. They usually have a good mechanical eye, and can tell at a glance whether a thing is straight, round, or square. Form is absolutely essential in a good mechanic, especially in working by the eye. When Form is the leading organ among the Perceptive group, it assists very much in remembering everything which it has seen, even in remembering words. In learning language, it will remember the form of words and characters. In its after-visions of what has been seen, it will always remember the form of everything.

SIZE.

Just below Individuality, and a little out from Form, is located the organ of *Size*. Its faculty determines the size of the objects which Individuality sees. It measures length, breadth, thickness, height, depth, dimensions, distances, &c.; compares objects with respect to size. It gives a good ability to remember the size of whatever Individuality perceives. Hence it gives a reliable judgment with respect to the size of men, horses, cattle, sheep, houses, farms, trees; all property valued in any way according to size. It measures distances correctly with the eye, and everything that pertains to size. Mechanics who have this large become very accurate judges of length, breadth, distances, size, &c.

WEIGHT.

Outward from Form, behind the eyebrow, is the cerebral organ of Weight, which exhibits the mental capacity for balancing, for walking firmly and gracefully, for maintaining an erect position, for supporting steadily the centre of gravity, and for judging of perpendiculars.

No man could walk were it not for Weight. Think for one moment of the ten thousand attitudes that the human body is thrown into every day; of the different burdens it has to bear—sometimes at its side, sometimes on its back, sometimes before it—and the thousand changes of the centre of gravity that every hour of active exertion produces, and then learn how essential is the faculty of Weight. It is strong in graceful and easy walkers; in good dancers, especially rope-dancers; in good horsemen; in sailors, generally, and men who go much upon buildings, and walk much in dangerous places. Weight is very essential in a quoit-player, in a sportsman, marksman; in a house-builder; yea, in almost every kind of mechanic. This is the faculty which the young child calls powerfully into action when he first begins to walk. While the two last organs which we have mentioned judge of the form and size of the things which Individuality sees, this judges of their weight, and, if in motion, their power of momentum.

Thus we have noticed the faculties which recognize three essential qualities of things.

COLOUR.

It will be remembered that every visible object possesses colour. Materiality is sparkling with an infinite variety of the most beautiful and dazzling colours. The light of the sun, moon, and stars is decomposed by a mysterious and wonderful power, which is hidden in almost every material substance, and is spread out in dazzling variety of colours, which make earth at some times a scene of the wildest enchantment. Behold the flowers of the wild prairie, coloured with every tint of the rainbow, multiform as the leaves of the forest, and as various in hue as in form; a sea of parti-coloured beauty! Behold the pearl of the ocean, the gem of the rock, the tree of the forest, the Aurora Borealis, the deep-blue sky, the morning dawn, the purple sunset! See Colour giving the chief beauty to each. Now, to recognize and be delighted with all this, the mind is endowed with a faculty to discern colours. The organ of Colour is located out from Weight, and nearly over the centre of the eye; a little out from the centre. Persons having it large are excellent judges of colour, lovers of colour, and everything possessed of beautiful colour, such as flowers, pearls, paintings, gorgeous clouds, gaudy dress, gaily-coloured birds, &c. They find a peculiar delight in gazing upon colours. They are always lovers of gay-coloured dresses, carriages, houses, furniture, and everything else that can have colour; while those in whom this organ is small have little ability to perceive the difference in varieties of colour, and find little delight in contemplating them. I think the founders of the sect called Quakers must have been sadly deficient in two faculties, Tune and Colour. They put an end to all music with them,—and I do not know but they would have stopped the throats of the singing birds, if they could have done it,—and clad themselves from head to foot in the almost colourless drab. They have always warred against Music and Colour. Perhaps all have not observed the fact that some persons discern colours much more readily than others; but all must have observed that some persons delight in colours much more that others. Some people are passionately fond of flowers and coloured paintings. They will spend half their time in cultivating flowers, and would be glad to spend the other half either in making or looking at paintings. Colour is absolutely necessary in a painter; otherwise he will not be able to colour his pictures properly, to dress them in their natural hues. A fancy-goods merchant ought to have this organ well-developed, and all persons who deal or practise in colours.

ORDER.

Long ago this great truth passed into a common saying : that "Order is Heaven's first law." If we observe everything closely throughout the vast field of nature, we shall see that it is conducted upon the principles of the most perfect order. The growth of plants and animals, the formation of crystals, the action of all chemical affinities, the gathering and movement of the clouds, the falling of the rain, the succession of day and night, and the seasons of the year, the movements of the oceanic tides and currents, the revolutions of the earth, and all the heavenly bodies, are upon the principles of the most orderly arrangements. Probably we cannot find anything in nature that is not done in perfect order. Sometimes we may not be able to discern that order; but everything is orderly, beyond a doubt. That man may shape his actions in harmony with nature, he is endowed with the faculty of Order. This, too, enables him to enjoy the contemplation of the grand scene of orderly magnificence about him. The idea of harmony, of completeness in arrangement, is given by the faculty of Order. Order in business, in labour, in study, in amusement, in everything, is absolutely necessary to the full attainment of the end in view, in each. There is no business or avocation in life in which Order is not necessary. Look at a farmer without Order. What a scene of confusion his farm exhibits ! His fields are all the same as one; his cattle are everywhere; his house, barn, and hog-pen are about the same as one; his tools are scattered on every part of his farm ; his work is all pressing him at once ; he does fifty things in a day, and does nothing after all. A mechanic, without Order is worse, if possible; he spends his whole time in getting ready for work. A merchant, without Order, is worse still; he has many goods in one box, which is a " salmagundi-box," containing a little of everything, where nothing can be found. A lawyer, without Order, is death to his clients ; for he never makes a plea, or a brief, in legal Order. A teacher, without Order, makes bedlam of a school. A preacher, without Order, makes crazy sermons ; not having either introduction, argument, conclusion, or exordium. Greatly to be pitied are the people who have to listen to such a preacher. But worst of all is a housekeeper without Order. If earth ever saw bedlam let loose—a picture of old Chaos, confusion personified, and an emblem of all misery—it is a house whose mistress has no Order. If any man on earth is to be pitied, it is the husband of such a woman,—unless he is as destitute of Order as she.

The grand idea of Order is, to have a place for everything, and everything in its place; so that it can lay its hand upon everything it possesses, whenever it is wanted, even in the dark. It has everything done by rule, at its proper place and time, and by its proper person. A man of perfectly cultivated Order will rise every morning at a certain time, breakfast at a certain hour, dine and sup at a fixed time, retire at an appointed moment, and do all his work with the same regularity. With him everything will go easy. He will always have his business done in time, and done as it should be done. He is punctual to every engagement, always ahead of his business, and accomplishes a great deal more than he who goes "helter-skelter" over everything. To a person who has large Order, nothing is more perplexing than disorder. Nothing will fret him quicker than to find things out of their places. If you wish to please him, never disturb his things. If you would do him a kindness, do it in order. If you would delight him, keep everything in order. If a man would accomplish much in his business, let him arrange it in the most perfect order. If a student would make rapid progress, let him arrange his studies, giving a certain time, and so much time, to each one, and abide by that arrangement. If the author would have his book interesting and profitable, let him make a good arrangement of his subjects and ideas, and present them in their proper order. If a lecturer or a preacher would be of service to his hearers, let him make a clear and natural arrangement of his thoughts. There is a natural order for everything, and everything should come in that order; and it is the office of this faculty so to cause everything to appear. There are some persons in whom Order is so large, in combination with a nervous temperament and large Combativeness, that they are extra nice, and excessively fretful, fault-finding, and unhappy at the slightest disorder.

The organ of this faculty is at the outer corners of the eye-brows. When large, it gives a sharp, angular appearance to the eye-brows.

CALCULATION, OR NUMBER.

Among the multiplicity of objects around him, man would be sadly deficient in ability to secure his own happiness, could he not count them, could he neither number the plants and animals around him nor the stars above him. Besides knowing the properties and qualities of objects, he needs to know the *number* also. How could he do his business with no ability to comprehend numbers, not even to count? It is evident that he could have no commerce, no science, and art could arrive at no perfection. Strike from man's mind his idea of Number, and he would go back to barbarism a thousand times sooner than he has come up from barbarism. We scarcely think, or can think, how much man owes to this single faculty. It is called into almost perpetual service in the every-day transactions of life. We count something almost every moment. It is the arithmetical faculty, the one that adds, subtracts, multiplies, and divides; the one that is used in the practical application of all arithmetical rules. It is used in every application of numbers.

When it is very strong, it will perform arithmetical calculations with wonderful rapidity. It is not unfrequently the case that we find arithmetical prodigies, individuals who will rush through the mental operations necessary to solve difficult questions with a rapidity that is past all calculation; and sometimes this power is strongly developed when the most of the other mental powers are very deficient. Instances of this kind have been presented to phrenologists, who have at once pronounced them "mathematical fools." They possessed great arithmetical powers, without common sense. An instance of this kind was found in a negro in one of the Southern States. He was *non compos mentis* in everything but Number. In this he was as remarkable for power as he was for want of it in other respects.

He would solve in his mind, in a very few moments, almost any question embracing only a simple combination of numbers. In all instances of this kind which have been presented to phrenologists, the organ of Number has been strongly developed, so that they have always spoken in most positive terms of this power. The organ is located outside of Order, at or above the outer corners of the eyes. When large, it gives width to the head at its location, and distance from the eye outward.

LOCALITY.

Just above and a little out from Individuality is found the organ of Locality. The faculty which it manifests is the mental geographer—the student of places, localities, situations, directions. It is the pilot-general of the traveller. It always gives directions about the *way*, keeps watch for the right road, knows the points of the compass, which way is home, and which way is the destined place. It is the faculty that never gets lost. It learns all about a strange city in a very short time, all its streets and byways; if strongly developed, will keep the points of compass in a wilderness. It loves to travel, delights in seeing new places, countries, and localities. It is this which often causes men to leave home, and friends, and all the pleasures of refined society, to roam through untrodden wildernesses and often amid the most imminent dangers. This is the exploring faculty. It gets up all exploring expeditions and companies. It is this, more than everything else, that opens new countries, discovers new continents and islands, strikes off into trackless regions of land and water, and makes bold adventures in wild and unknown parts. This was, doubtless, strong in Columbus and those who sailed with him in his great enterprises. In navigators, voyagers, pilots, explorers, travellers, and others, it is always large. I imagine there is a prodigious development of it in the head of Colonel Fremont. I have in my mind an idea of his forehead, and I have never seen him nor a picture of him, nor a description of him, nor anything, but what little I know of his life, from which to judge. I imagine that the Perceptives are much more strongly developed in his head than the Reflectives. At any rate, his Perceptives must be very strong, with a remarkable development of Locality. I imagine that he has heavy, arched brows, with a somewhat retreating forehead. This was especially true of Audubon.

Locality is very necessary in geographers and surveyors; it is also in phrenologists. It is necessary to locate the organs. It is generally large in hunters, and all wanderers into unknown parts. It gives an excellent memory of places. It daguerreotypes them, as it were, in the soul. An image of them is always retained. It frequently assists very much in the memory of words, sentences, numbers, and anything on a printed page, by remembering its place on the page. Many people learn to repeat verbatim by this organ. It always has an exact picture of the page before it.

The intellectual organs thus examined are called Perceptives, because it is their office to perceive some individual quality or property. Individuality perceives objects; Form recognizes their form; Size, their relative dimensions and distances; Weight exercises its peculiar function with respect to these; Colour observes their hues; Order, their arrangements; Calculation, their numbers; and Locality, their places. Thus we find that the faculties which observe the various qualities and positions of individual objects are situated in one group in the most frontal portion of the brain. These are the observing faculties, and always give keensightedness to a mind—readiness to observe the peculiar aspects of everything they see. They give a *practical* turn to the character, and when very large in any man, make him a *practical* man. They are the faculties that go farther in making up the artist than anything else. They are necessary to every artist; Size and Form, especially, to the sculptor; Size, Form, and Colour, to the painter. They are almost absolutely necessary to a practical observation of all scientific pursuits. The practical zoologist, ornithologist, botanist, mineralogist, geologist, chemist, and geographer must have them in very active development, or they cannot succeed. Herschel, the great astronomer, and Humboldt and Cuvier, the two greatest naturalists in their departments, had very

strong developments of the Perceptives. These are absolutely necessary to gather up that information about natural
objects, to gain possession of those facts which are the premises
from which all scientific conclusions are drawn. These powers
must gather up all needed facts before reason can apply her
power in deducing her grand conclusions, forming her general-
isations, and establishing her well-built theories of the natural
laws and operations.

EVENTUALITY.

But we should suppose that these busy fact-gatherers would
need some general store-house, or record-keeper, or historian, to
keep a true account of the myraid observations made by them ;
and so they have. The results of their observations are handed
up to Eventuality, whose location is just above them, but in
close proximity, in the centre of the forehead, and whose office
is to keep a faithful record of all that passes before the
Perceptives. The stirring events that the Perceptives behold
are noted by this event-keeper, and treasured up for after use.
It is frequently remarked of individuals, that they have or have
not good memories. Persons of large Eventuality generally
possess a good memory of events, and hence are good story-
tellers, historians, and are generally walking histories of everything that has passed before them during their lives. But the
term " memory " is very indefinite in its meaning. In truth,
there are as many kinds of memories as there are mental
faculties. Each faculty remembers the objects upon which it
fixes its affections or attentions. Thus, Ideality remembers
beautiful things ; Sublimity, objects of grandeur ; Constructiveness remembers machinery ; Acquisitiveness, money ; Alimentiveness, good dinners ; Calculation, number ; Size, Form,
Weight, Colour, the several peculiarities which they observe ;
and so on to the end of the chapter of man's natural endowments. Each one has a memory of its own. It is the office of
the faculty now under consideration to observe and remember
events ; hence it may be called the mental historian, the record-
book of the soul. As such, it should be strong in every mind.
It makes matter-of-fact men ; furnishes materials for reflection,

objects for thought, meditation, and study, and is absolutely essential to a sound mind, to a philosopher, to a well-balanced intellect.

Without good Eventuality, the reflective intellect has but little material upon which to use its energies. The reflective powers are compelled to go to Eventuality for the great mass of the facts which it is necessary for them to use in their arguments, processes of reasoning, forming their theories, philosophies, judgments, &c. " From what can we reason but from what we *know?*" asks a poet and a philosopher. Eventuality is the *knowing* faculty ; the Perceptives are the *seeing* faculties ; and the Reflectives are the reasoning faculties. Thus the Perceptives go ahead and look up and search out things and facts, observe all their various particulars in the minutest detail; Eventuality observes and records all the facts, and events, and material circumstances which are thus gathered up, and hands them up whenever they are needed to the Reflectives, which receive them, arrange them into their proper classes and orders, study their various causes and results, and combine them to form sciences, theories, arguments, philosophies, and reason from them according to the true method of induction. Thus it will be seen at once, that in order that a mind may be powerful, it should have these three classes of faculties in equal vigour and activity. If either part is too strong, it will be too much biased by the actions of that part.

Let us now attend to the Reflective faculties.

CAUSALITY.

The central and predominant power of reflection is Causality. It is, strictly, the reasoning power ; that which traces the relations of cause and effect ; which deduces conclusions from first principles ; which finds the logical deduction from given premises; which reasons from the seen to the unseen, from the known to the unknown ; which, from the operations of nature, reads her hidden laws ; in the works of God sees the evidences of His being and nature; from the outward actions of man determines the secret workings of his heart. It is the power, that with almost angel ken, penetrates the veil of the visible world, and gazes upon the mysterious springs of motion, life, and being that lie below. It is the architect of those pyramids of thought which stand like majestic piles of light along the road of human progress, thick as the stars on the brow of night, and which men have called sciences. It gave being to the thousand philosophies that have risen in the ages past and present, to account for what is, and accomplish what should be. It is the maker of syllogisms ; it is the founder of arguments ; it is the drawer of conclusions. It is the inquiring, investigating faculty. When large in children, it causes them to ask a thousand questions, questions which oftentimes would puzzle a philosopher. It is always anxious to know the "why and the wherefore," the reason for everything. It is strong in the philosopher, the statesman, the true logician. No man ever became truly great in any of the solid sciences or any of the learned professions without a strong development of this mental power. When strong, it is not apt to be brilliant in its displays, but solid ; not quick, but sure. It lays the foundation of true greatness, not of dazzling splendour. When it is strong in a harmonious combination with the other mental faculties, and a strong and mental temperament, it makes a man of enduring greatness, and crowns him with an immortal name. But in order that it may exercise its true functions, and give to the whole mind its true dignity and glory, that mind should be favourably developed in every particular. It gets all its premises, its facts, its known things from which to reason, at the hands of the other faculties. It has no other way of getting them. If its facts are incomplete, or but partially given, or falsely coloured, or half conceived, its premises will be false ; and its conclusion must necessarily be false, if its reasonings are correct. So the power, and beauty, and glory of the faculty depends in a great measure upon the harmonious combinations, with which it is found connected. Its locality is about an inch from the centre of the forehead above and out from Eventuality. When large, it gives prominence to the forehead ; when very large, a high, perpendicular, and expansive forehead.

COMPARISON.

Comparison is the second of the reflective organs. It is located in the centre of the forehead, above Eventuality and between the two organs of Causality. When large, it gives prominence to the middle of the upper part of the forehead. It is the servant of Causality, and assists it both in seeking the truth and in explaining it to others. Its office is to see analogies or likenesses. It demonstrates or explains the *unseen* by reference to the *seen*. It compares the unknown with the known, and in this way seeks and explains the operations of the invisible world of truth, principle, and law. Causality often calls upon this faculty to explain or illustrate what it finds difficult to demonstrate. This is the faculty in which have originated all fables, parables, figures, tropes, comparisons, &c., with which all human language, both written and spoken, abounds. In the earlier ages of language and civilisation, it was used almost perpetually. The language of savage and barbarian nations originates almost wholly in this power of mind. It is a simple comparison of ideas with things. This faculty inspired much of the language of the Old Testament, and not a little of the New. This and Sublimity assisted largely to the conception and expression of the Apocalypse. Much of the reasoning of minds which have it large, is done by analogy. He who has it large always explains all his ideas by comparison with something else. It uses the word "like," "as," "as when," very often, as suggestive of its figures. Let it not be forgotten that its use is to see *likenesses* or *analogies*.

MIRTHFULNESS.

On the other side of Causality is another organ, the servant of Causality, the office of which is to perceive *differences* or incongruities. It is quick to perceive absurdities, is always on the look-out for all incongruous, inharmonious, inconsistent things and ideas; and often so ludicrous are its conceptions that it sets every member of the mental family into a roar of laughter. It is the maker of wit, fun, humour, sport, merriment, mirth, and all the family of real laughables. When this is joined with Combativeness and Destruction, it makes ridicule, sarcasm, venom-toothed sallies of wit, and that whole family of despicable serpents which bite while they laugh. It is often very useful in demonstrating the falsity and absurdity of a position. Not everything is witty that is laughable. Things are laughable sometimes because they are destitute of wit. A thousand things please us that are destitute of wit; and some witty things do not make us laugh. He who has this organ large is always a dear lover of fun, and goes through life in a whirl of wit and merriment.

LANGUAGE.

One more organ completes the Intellectual group, and that is the servant of the whole mind. It is Language. It is located above and back of the eyes. When large, it presses the eye out and down, so as to give it the appearance of fulness. To determine the real strength of Language, the distance of the eye from the ear must be measured. It is the object of Language to give expression to the thoughts and emotions of the mental group. When it is strong and active, it gives great fluency in the use of language, readiness in expressing every shade of thought, ease and gracefulness to expression. If very large, it gives great verbosity and redundancy of language. It does not always make great talkers, but always gives a free, easy, graceful use of language, and a readiness in learning language. Small Language finds difficulty in expressing its thoughts, staggers and stammers at words, makes bad selections of words, and often fails in expressing the true idea. Good Language adds greatly to the expression of vigour and beauty, and adds much to its possessor's interest in society, as well as his usefulness. Perhaps no faculty gives a higher charm to cultivated minds than this. It clothes ideas, already sound and beautiful, in the sparkling habiliments which alone would become them well. It robes the expressions of the affectionate nature in words soft as love's own balmy breath, and musical as the notes of old "sweet home." It carves the thoughts of the intellect into statues of living beauty, and paints the aspirations of the moral powers in the rainbow hues of living life and hope. So chaste, so delicate, so refined and clear are its vestal robes of light with which it arrays the ideal offsprings of mind, that it gives to conversation, to literature, to life, a pure and elevated charm, as well as strength and brilliancy, which is the last touch of cultivated refinement.

What though your thoughts are rich and rare ; what though your emotions are strong and high ; what though your loves are like the angels'? If you give them not appropriate expression, they fall upon the mind stale and powerless ; they kindle no kindred fires nor wake other souls to vigorous and impulsive thoughts. There is real poetry in nearly all souls ; but few only are gifted with the power of its expression. Who that reads a great poet, thinks not, feels not, with him ? But how few are they who can *express* what he has. In the hand of powerful Language, human words and forms of expression shrink and expand, rise and fall, bend and yield, with an infinite elasticity, to express the finest shades of thought, and the most delicate tints of emotion which take shape and form in the human soul.

When strong. Language is truly cultivated, it makes everything its servant. The stars, the flowers, the rocks, the waters, the clouds, oceans, days, nights, dews, showers, frosts, all things speak for it and give variety, form, and force to its multiform modes of expression. There is scarcely anything that strong Language may not and will not use with propriety and power in the utterance of the thoughts it is required to bear out to the world. Then how proper, how useful, is the cultivation of Language. Long has written language been an academic study. The scholar has for centuries been searching among its living forms and fossil relics for its mines of wealth and power. Some have spoken against this faithful and perpetual study of language, both living and dead ; but they surely have spoken unwisely. The mental faculty we are considering bids us cultivate it long and well. Its office calls us to the work early and late. This faculty is the servant of the whole mind. The passions, the affections, the imagination, the reason, the moral sense, each individual power of the mental man, is calling daily, and almost hourly, for an expression of some burning passion, feeling, or sense, or careering thought or emotion. And how can all this be chastely, properly done, without a due cultivation of the power whose office it is to do this work? Says an inspired penman, "Words fitly spoken are like apples of gold set in pictures of silver."

LECTURE X.

The Moral Sentiments, the crowning excellence of the Nature of Man—Nobility of Moral Excellence—Phrenology harmonises with the Bible—Benevolence: its Location—Benevolence, God's Image in Man—Benevolence, the Soul's Good Samaritan—Melancthon, Oberlin, Howard, Josephine, Mrs. Fry—The Spirit of Forgiveness—Benevolence, the Prince of Peace—Veneration, the Central Element of the Soul—The Basis of Religion—Communion with God—Its Cultivation and Abuse—Spirituality—Credulity—Faith—Hope—Pleasures of Hope—Conscientiousness—Its Godlike Moral Power—Firmness—Beauties of Mental Science—Closing Appeal to study Phrenology.

WE come now, my young friends, to the imperial crown of man—the Moral Sentiments. If, in our passage through the mental apartments, we have been delighted with the beauty and utility of every part, we ought to be enchanted with what we now behold, yea, we ought to feel like bowing in reverent respect before the august powers that sit in regal majesty in the throne of mind. Here is a sinless, holy circle of the servants of the living God, doing his will, honouring his law, and adoring his being.

Whatever is amiable in human sympathy, beautiful in benevolence and philanthropy, grand in forgiveness; whatever is holy in purity, noble in righteousness, excellent and inspiring in faith, or sublime in religion, is found to issue from this birth-place of all moral good, this cradling-place of human excellence. Are there men of such sterling integrity that they command our respect, that they make the deceiver feel rebuked in their presence—men whom a thousand thrones could not buy from duty, whom wealth cannot corrupt, or earth seduce; are there those whose presence is a living charm of goodness, who invite mankind to love and charity by a voice above words, a spirit-voice that whispers from soul to soul the noble utterings of self-sacrificing beneficence; are there those who seem to dwell on the verge of a better world, within the sight and influence of heaven itself; are there earth-angels about us, ministering to poor, frail humanity, saving it from utter ruin, and pointing it to a holier clime, a purer and nobler state? They are made what they are by their strong endowment of the moral sentiments. There is something dazzling and grand in intellectual greatness, in the coruscations of genius, in the giant exertions of reason, and the bewildering strides of imagination, something startling and even sublime in the rapid movements of the mighty general as he moves his countless . hosts of mail-clad warriors among the nations, bidding them bow to his nod, live on his breath, and grow strong on his supplies,—in the dictates of the emperor whom millions obey, in the words of the philosopher which breathe a quickening life over the whole world. But there is something still nobler and grander in the lofty purpose, in the noble daring, in the godlike intrepidity of a truly good man. There is grandeur in ruling nations, but there is a sublimer grandeur in ruling oneself. He who is greatest in goodness is most an angel, nearest to God, the noblest earthly being that lives.

All that makes this greatness, comes from the moral sentiments. They are made and given to rule the mind. They govern by a divine right. They wear the crown because they alone are worthy. Their authority, it is true, is not always obeyed. Their crown is often trampled in the dust by the brutal outrage of the propensities. And this is just what ails our crazy world. It obeys not its God-appointed authorities. It wars against the just rule of its only just rulers. It listens not to the voice of wisdom from the mental throne. And so long as this state of things remains, this will be a wicked, miserable world. It cannot be otherwise. And this is one of the great lessons of Phrenology. It bids us, as individuals, to remember and revere the divine authority of the moral sentiments. When we learn this lesson, well will it be for us. Here, as everywhere, Phrenology harmonises with the Bible. Divine teachers tell us that charity and righteousness must reign in the heart, or man cannot reach the end of being; they tell us that the principles of moral purity and perfection will alone lift the soul from the vale of sin and tears, into the upper light of the kingdom of heaven; they place, as the grand end of all our efforts and prayers, the attainment of that state of mind in which truth and love shall rule all the powers, guard against every temptation, and lead the soul into the beautiful haven of peace.

The great end of all moral and religious instruction is, in reality, to give authority and vigour to the moral sentiments, to make them in reality what they are of right—the directors of the human powers. For this revelation was given; for this Jesus came, and other spiritual teachers; for this the Church was established; for this its ordinances are perpetuated; to this end all great moral reforms look; to attain this end moral essays and lectures are read all over the civilised world; to this end the gospel minister directs all his labours, and the true teacher pursues it with a steady care.

When the moral powers rule, peace reigns in the mental world; when they are made subjects, war and discord prevail. All that we love most in this life, and prize most in the future, comes to us by obedience to the properly constituted authorities of the mind. Let us, then, as we value good, as we love excellence, as we hope for blessedness in the future, examine these powers and learn, if possible, their true use and beauty.

BENEVOLENCE.

As we leave the intellectual region of the cerebrum, and pass up and back, we meet at once with the organ of Benevolence, a large and powerful organ, located on the top of the front head, showing at once its dignity and office. It is evidently placed there for a noble purpose, and it must manifest an exalted faculty of mind. As we ascend from the base of the brain, the organs become more elevated in office as well as in position. Here we have risen to the top of the frontal lobe of the brain, and are contemplating the highest power conferred upon mankind—the power of goodness. Goodness we have always been taught to regard as one of the most admirable attributes of the Divine Being. Goodness is really a part of the divine character. Goodness is, so to speak, also an attribute of man. It constitutes a part of his nature. It is a portion of the divine image in man. Moses says that when God had finished the present order of creation, the last and crowning work of which was man, he pronounced it " very good." And a proper eulogium was this upon man, made in the likeness of his Maker. The name by which this faculty is generally known is Benevolence. It is the good Samaritan in the soul, which goes about doing good. It is that faculty which inspires all the kindly feelings of the human soul of a general nature, all humane emotions, all sympathetic tenderness for the suffering, the sick, the needy, the lost, the sorrowing, and the desponding.

A mendicant begs bread at your door. You look upon him, a poor, destitute, forlorn, unhappy creature. You think what reverses of fortune he must have known, what woes he must have experienced, what scenes of wretchedness he must have passed through, what coldness he must have met at the hands of his fellows, to have brought him down from the high eminence of a noble, independent man, to that of a degraded mendicant. A thrill of pity, and a sense of sorrow runs through your heart; a feeling of almost maternal softness steals over you; a deep desire to relieve the sorrows of the " poor, wayfaring man," succeeds. All this tenderness is the action of Benevolence. It comes from this spring-source of goodness within. And if you can look upon such a man and not feel thus, you may know that you yourself ought to be such a mendicant, begging the bread of moral life at the hands of your God.

The story of the " good Samaritan," told in the Gospels, is one of the most touching exemplifications of human Benevolence

ever exhibited to mortals. It is so full of the genuine spirit
of charity, which is so well supported by corresponding outward
acts, that it is really nothing more nor less than an exact
personification of the feeling of Benevolence, in contrast with
its destitution. No human artist has ever drawn so perfect a
picture of this glorious, heavenly power. The whole life of
Jesus is one splendid illustration of the principle of Benevolence.
It is spread out before us as one magnificent plain of charity,
glittering all over with the flower-gems of this image of God.
But in this great plain, resplendent as the arching heavens,
there are star-piles of peculiar glory, which tower up in radiant
grandeur here and there, like pyramids of sapphire, or mountains
of polished diamonds. At Nain, at Bethany, at the time when
the wicked woman was brought to Him in the Judgment Hall,
and on the cross, He gave us the sublimest exhibitions of benevolence. This last is the eclipsing sun of moral grandeur. In
this the power of benevolence shines in unclouded majesty.

It is exhibited in less brilliancy in the lives of all good men.
Melancthon, the associate of Luther in the Reformation of the
fifteenth century, gave us its beautiful features in his life, all
the more lovely to behold because presented in such striking
contrast with the surrounding hatred and malice. That age
was a boiling ocean of animosity. The elements of revolution
were powerful, and in powerful action. Very little of the milk
of human kindness tempered the cup of gall drank by the people
of that age. Hence, as we look upon that boisterous sea of
human passion, the benevolent spirit of Melancthon appears
almost like a Jesus walking upon the raging waters, saying,
" Peace ; be still."

The life of Oberlin is a thread of benevolent LIGHT, winding
through the darkness of that pall of blackness that shrouded
the people of his day.

The track of Howard is to this day glittering with kindly
radiance. He made the prisons of Europe his home, the
wretched prisoners his companions, his brothers ; he spent his
life and an immense fortune in doing good to the poorest and
vilest of our race. For this he is justly remembered as a mes-

senger of goodness, and his eulogies have become the pæan of praise of the civilised world to the genius of Benevolence.

Josephine, the first wife of Napoleon, should never be forgotten when speaking of benevolence. It was the pride of her heart to say, that in all her great and stirring life, as fraught with mighty actions and checkered with reverses as any woman's that ever lived, "she never caused a tear to flow," or failed to dry the eye that wept, when it was within her power.

Mrs. Fry, the friend of the degraded, whose mission-path was among the vile and the low, who felt herself commissioned of the Most High to do good, demands a tribute from all the benevolent.

In less noted lives this noble feeling has been strongly shown ; and in characters unknown to fame, it has burned like a pure altar-flame upon the shrine of goodness.

Wherever good has been done, wherever the hand of charity has been extended to relieve the wants of the human kind, wherever self has been sacrificed to bless others, there Benevolence has made her angelic plea, her voice has been heard, her spirit prevailed.

Would you know what she has done? You must read the record-book of God. Her noblest deeds are often done where no eye but God's can see them. But still the light of her glorious achievements shines brilliantly in the earth. Behold the asylums for the poor, the unfortunate, and the vile, which are erected everywhere, in civilised society. Behold the hospitals and charity-schools, the benevolent associations, that are founded in every town. Almost every crying sin and great misfortune has some benevolent association designed to relieve those who are its victims. All these speak forth the praises of Benevolence.

But, really, the great field of Benevolence is in the every-day life of the masses of men. In the homes and daily actions of our fellows we witness its most beautiful and constant actions. In the acts of kindness, in the words of charity, in the smiles of beneficence, in the watchings, and toils, and labours of men for each other, in the peaceful flow of life's duties and cares, in the harmony of nations, in the peace of neigbbourhoods, in the union and happiness of families and associations, we witness the guardian spirit of Benevolence. Benevolence must temper and sweeten all our affections or they will lack that steady flow, that kind and forgiving spirit which is always necessary toward the objects of our love. Now love, in its ordinary sense, is of all things the most exacting. It will submit to no cessation, no turning of the eyes away for one moment, and is very likely to find a thousand causes of jealousy in the lives of the calmest and most constant lovers, if that love is not steadied and strengthened by Benevolence. Benevolence always enshrines within itself the spirit of forgiveness, the noblest virtue that crowns the lives of mortals. There are some persons so constantly tender and kind, so completely self-sacrificing, so wholly devoted to the happiness of others, so ready to yield to the wishes of others, so forgetting and forgiving, that we involuntarily name them "human angels." They walk before us as guardian spirits. They breathe about us an atmosphere of peace and love. They hallow life, and make it dear. Their hands are soft to the sick, their words are gentle to the erring, their smiles are always upon the desponding, and their sympathies with the suffering. We love them, and cannot help it. These are they who are largely endowed with the faculty of Benevolence, and if you will observe, you will find a great elevation on the frontal part of the upper head.

It is the office of Benevolence, in particular, to walk upon the sea of passion when the storm is up, saying, "Peace; be still." It is its office to smile in the face of anger and hatred; to entreat with the warring elements to be at peace. It is, in fact, the universal peace-maker and peace-keeper; and all the blessings that come through peace are due to Benevolence. Then how earnestly, how faithfully should it be cultivated. Its voice should always be heard, and we should ever be striving to see how much good we can do to others. Instead of studying always to see how much we can get from others, it should be our constant study, the grand aim of our lives, the promi-

nent object for which we live, to see how much *good we can do.*
This is the voice of Benevolence.

If you would see the best description of this feeling that has
ever been written in human speech, the most perfect descriptive
elucidation of its beauty and excellence, turn to the thirteenth
chapter of First Corinthians. You will find it there as a master-
piece of literary skill. Read it long and well. Remember it,
and let its spirit find a dwelling-place in your hearts.

VENERATION.

Back from Benevolence, and directly in the centre of the top
of the head, is located the organ of Veneration, the primary
office of which is to give expression to the worshipping faculty
of the human soul. It gives the original idea of the Supreme
Being, and places us before Him in the attitude of worshippers.
It is emphatically the religious faculty. It is the central organ
in the grand moral crown of man—the central, noblest, holiest
power of the soul. It is that moral link that binds man more
closely with his God; that spiritual garden where the creature
walks in sweet companionship with his Maker; that feeling
which adores, worships, loves the Divine Being, and which
clings to Him with a holy, a devout, and reverential affection.
This is the central, all-radiant sun-jewel set in the crown of the
soul. Its light is pure, tranquilising, and spiritualising. The
sentiment of veneration, of worship, of love to the Divine Being
is the highest and most sublime of any that man is capable of
cherishing. It has for its object the perfections of the god-
head. It fixes its regards upon the immortal glories of the
great Father of Lights. It binds itself to a Being, fitted as no
other being is, to impart to the soul the highest moral grandeur
that created beings can attain. It communes with the omnipotent Spirit of Love, which transfuses its energies through
the wide creation. It is the upper window of the soul, which
opens into the clear, radiant light of God's eternal home. It
is the ladder of Jacob, on which angels ascend and descend in
intimate intercourse between the soul and its God. It is an
affection, a love, as positive, as real, as warm, as imperative in

its demands for activity as any implanted in our natures. It is the grandest and noblest affection of the soul, because it fixes its regards upon the sublimest and holiest object in the universe. And its influence in the mind is more salutary and holy than any other, because of the strength of the feeling and the nature of the Being upon which its adoration is fixed. As God is holier, lovelier than any other being, the affection for Him is more excellent in its influence upon the mind than any other.

No mind can be perfect, no other affection can rise to its highest degree of perfection, no faculty to its most exalted state without the sanctifying power of this sentiment. If we would perfect our natures, if we would exalt our affections, if we would purify our souls, if we would reach the acme of true human greatness, we must give to the sentiment of Veneration its full and proper influence in our minds. It is the basis of religion ; the religious impulse which has spread its influence and its testimonials throughout the world. It is opposed to all evil, opposed to the undue exercise of any and every faculty. Its will is the will of God, so far as it knows the Divine desire. It is opposed to all things which militate against the laws and precepts of the Most High. It loves obedience to God. It delights in dependence upon Him. It recognizes His hand in every created thing. It feels Him everywhere, and rejoices in the feeling. It offers praise and thanksgiving. It lifts itself in prayer. It bows itself in worship. It venerates God and all things kindred with Him. It loves holiness, loves purity, godliness in thought and life ; loves devotedness to truth and right ; loves sincerity, sanctity of spirit ; loves goodness, humility, meekness, love. In fine, it loves all things which are kindred with God, and upon which God smiles.

This is the real function of Veneration. It fills the throne of dominion within. And this being its office, it should be one of the primary and principal objects of all persons, and especially the young, to cultivate it well, to direct its energies to the one living and true God, to inspire it with all possible power, to enkindle all its holy fires, and spread its sanctifying charm through all the faculties of the soul. Its natural language is praise and prayer. It delights to make known its love to the one object of its regard. In this respect it is like any other affection. All affection is communicative, and delights, above everything else, to make itself known to the object on which its fixes its attention. It delights in praises of that object, in expressions of respect and attachment. It never wearies in imparting itself, in making known the depth and strength of the fires within. Hence praise and prayer are the natural language of Veneration. These inspire it with power and activity. These enkindle its fires. These give it cultivation. Religious worship augments its activities; sanctuary services awaken its powers; religious associations give permanency to its feelings. Hence public and private worship, religious meetings, and exercises of all kinds are profitable for the cultivation of this the highest faculty of the human soul. All religious ceremonies and exercises were established and are supported by this sentiment. They are the visible expressions of its office and power. Their atmosphere is the element in which it delights to live. So we see that religion, worship, praise, prayer, are as natural as they are revealed. All this is the voice of this sentiment, speaking out its own nature, and reminding us of its great Author.

I have not time to speak more upon this faculty. But permit me to urge its cultivation upon my young friends, as one of the most sacred duties of life. Neglect it not. Neglect not the sanctuary; neglect not religious reading, religious reflection, the formation of religious opinions, and the cultivation of a religious life. The highest beauties of your souls, the finishing touch of your characters, the sweetest charm of your lives will be given by due attention to this your first and last duty.

It may not be improper to remark that even this excellent faculty is liable to great abuses. When exercised without due enlightenment, it exercises an unhappy tyranny in the mind and is the origin of bigotry, intolerance, superstition, and all kindred vices. It is not an independent power. It requires the assistance of the intellect, the softening charm of Benevolence, and the mild influence of the social affections, to give it its highest power and its noblest office.

SPIRITUALITY.

On each side of the organ of Veneration is located that of Spirituality. It manifests that faculty in man which contemplates his spiritual relations, which recognizes spiritual existences, spiritual life, labours, and joys. Man is everywhere a believer in unseen realities, a believer in spirit. It is this faculty which constitutes him such.

It is this which gives him an idea of spirit, without which he could not think of a spiritual being, or have any conception of a spiritual life. And inasmuch as every faculty has in existence the proper object of its contemplation, and the existence of the faculty proves the existence of the object, the existence of the faculty of Spirituality proves the existence of a spiritual life and of spiritual beings. So the existence of Veneration proves the existence of God, its proper object. Thus, from man's very mental nature we get the elements of religion.

And it may here be remarked, that the religion of his nature is exactly the religion of Revelation, the religion of the Spirit. He, then, who opposes religion, opposes the plainest teachings of his own nature, turns a deaf ear to the voice of his own soul. Men will have a religion as surely as they will love, or reason, or eat, or sing. They are made for religion as much as for any of these. The sentiments of Veneration and Spirituality are as purely and positively religious as love is affectionate. Being, too, the central and highest organs, they must manifest the central and highest faculties of the soul, and exercise the noblest authority and the purest influence. To them, then, should be given the most thorough cultivation and perpetual pre-eminence.

Spirituality is truly the prophet-seer of the soul, and it is through this organ that the grand truths of Revelation have been made known to man. This is the entrance-window of spiritual light, the visiting-ground of angels, the communion-table of spirits. This is the door that opens into spiritual life and hence, when it is strong, spiritual influences are very strong in the mind, and spiritual subjects are contemplated with the most intense interest. It then confers upon the mind a readiness to believe in spiritual presences, and to credit pretended revelations from the spirit-world. It gives the feeling of the nearness of spiritual beings, and of the actual presence of the spirit-world. Hence, he who has this organ largely developed is very likely to believe that we are surrounded with spirits, that guardian angels attend us, and often whisper in our souls of coming events, and give us intuitive impressions of important truths. And hence such persons are likely to live as though in the presence of angels ; to live pure, holy, and consecrated lives. When they have otherwise well-balanced minds, they are, indeed, our noblest, our highest, our purest human souls. They are almost spiritualised already. They live much as spirits live, feel much as spirits feel, and enjoy much and intensely their communings with the great Father of spirits, who is Himself a spirit, and who seeks such to worship Him. To me such souls are supremely beautiful, congenial, and dear. I love them as by intuition, and cannot help feeling that we exist in a sort of mystical oneness, or spiritual union ; which, perhaps, is a faint image of that referred to by Jesus, when He said, " The Father and I are one."

When this organ is small, it gives the opposite state of mind upon this subject. It is usually small in sceptics, unbelievers in Revelation, infidels, atheists, &c. It is unfortunate to have a small development of this organ, and that of Veneration ; for these are the natural and most powerful resisters of material and carnal tendencies ; they hold the strongest authority over the propensities. When they are small, the propensities generally rule the character. Either the appetites, the passions, or the lusts, will generally be too dominant, and greatly mar the character, if these religious sentiments exert but a weak influence.

They may be too strong, or too strongly excited. In that case, they produce the most disastrous results upon the mind. But these sad results, I believe, are always occasioned by false appliances, false and unnatural stimulants, which appeal to the fears as much as to the religious sentiments. Religious truth will never disturb the mind, however strong may be its religious tendencies. It is error that bewilders; truth makes clear.

Whoever would adorn and elevate his mind, whoever would perfect and beautify his character, let him cultivate much, and with great care, the religious faculties of his mind.

HOPE.

We come next to the faculty of Hope, about which poets have written, minstrels have sung, and lovers have talked so much. It is life's sweet charmer; earth's gladdener; the index of perpetual sunshine; the wreather of the heart in flowers. It is the great holiday-maker, the jubilee-singer of time.

I need not stop to talk to you about the office of Hope, nor of the darkness, gross and deep, that would crown the world were it not for it. Read Campbell's "Pleasures of Hope," and you will be sufficiently enlightened upon this point. It gives, when large, a cheerful, happy, hoping, castle-building, good-time-seeing disposition; a fearless, gladsome, merry heart; a soul alive with high expectations and glorious aspirations. When small, it gives the very reverse of all this, a down-in-the-mouth, desponding, deploring, hopeless, gloomy cast of mind. Pity, oh, pity the hopeless man! The blue spirits of evil are about him half his time. The sky of his soul is canopied with dark foreboding clouds. Despair often wraps his spirit about with her folds of blackness. He lives in gloom, and dies in fear. Shadows of good are to him spectres of evil. Large Hope gives to one's life, philosophy, religion, and character a cheerful aspect; and as it is associated with the religious sentiments, it exerts, and was designed to exert, a powerful religious influence. United with Spirituality, it forms *faith*. Large Hope and Spirituality give a strong and happy faith. Hope believes in a full redemption; trusts implicitly in the goodness of the Divine Being; believes all will be well with mankind; sees the future clothed in the radiance of perpetual day; rejoices in the full prospect of immortal felicity; and sings a song "of joy unspeakable and full of glory," as a present prelibation of the draughts of life it will hereafter quaff.

Clear wings and open sailing should be given to Hope in every wind. Hope is natural and beautiful in the character; despair is monstrous and unsightly, and should never be permitted. The organ of Hope is located at the back of Spirituality, and above Sublimity. It exercises a marked influence in the character, and should always be carefully cultivated.

CONSCIENTIOUSNESS.

Back of Hope, and above Cautiousness, is Conscientiousness, the love of truth and right. It is the spring-source of integrity. It has been said that "an honest man is the noblest work of God." It is the inspiration of this sentiment that makes him such—which crowns him with his real nobility. A great writer has said that "the two most beautiful things in the universe are the starry heavens and the sentiment of duty in the soul," —a sentiment most noble and true. If there is a being in the government of the Most High who is worthy of the heart's esteem and high respect—if there is one to whom we should bow in willing reverence, and in whose presence we should feel as though by the side of an angel, who should awe us while he secures our love—it is he who has a strong and ruling sense of duty in his heart. It appears to me as though there is something of God in this feeling. It works a divine inspiration upon me. It fills my soul with heavenly images, and binds my heart to its possessor. It has a ravishing charm, and works as though by miracle upon my inner senses.

This sense of duty in the heart is inspired by Conscientiousness. The ultimate of the authority and office of this sentiment is to impart this sense of duty. The ideas of obligation, of responsibility, of faithfulness to trusts, of rectitude, of justice, of right, are conferred by this faculty. The voice of this sentiment is for right. It has but one law written in the heart of its being, and that is the law of right. It is a stern, noble representative in man, of the attribute of justice in God. It is communicative like other feelings, and desires to impart itself

L

to others. It wishes to inspire its own glorious spirit everywhere, and make all hearts redolent of its light, and the sanctity of heaven. It suffers in the presence of wrong, sorrows at injustice, weeps when any creature fails in duty. How much of sanctity, of holiness, of god-like moral power this faculty, when strong, imparts to the human soul! It is the citadel of moral force, and should be guarded well. Faithfully should it be cultivated. Nothing should prevent a thorough and perpetual cultivation of this right arm of all morality. The present and everlasting interests of everyone depend upon the cultivated energies and activities of this sentiment. Then injure it never, oppose it never, outrage it never, question never its teachings ; be true to its voice, heed its warnings, obey its dictates, walk by its counsels, comply with the letter and spirit of its law. Come what may, frown who will, hearken to the voice of duty. It is God in the soul, speaking a language beautiful as the words of heaven. Its language is, "*Fiat justitia ruat cœlum*"—Let justice be done though the heavens fall.

. F I R M N E S S.

Above Conscientiousness, and in the very highest region of the back head, is the organ of Firmness. It manifests the faculty of unchangeability, or steadiness to all purposes, in man. It is the human image of God's immutability. It crowns the character with that glorious virtue which no poet or divine hath yet sufficiently praised—that virtue which is "as the arching heavens for beauty," and "as the pillars of earth for firmness." I mean *fidelity*. It is fidelity that crowns every virtue ; fidelity that spans the moral world with the arch of beauty.

When this faculty is strong, it gives great permanency and steadiness to the character. It holds it as with reins of iron. It makes a character as reliable as the mountain pillars. Its opinions, its feelings, its habits, are almost as changeless as fate. It is the chief ingredient of that great virtue perseverance, without which nothing would be accomplished. It is strong in all great characters. It has done much to make them great, by making them steady to their purposes. When it is small, the character is feeble, changeable, fickle as the wind, unstable as the clouds. Strong Firmness gives dignity and moral grandeur to the character, and should be possessed by all minds.

We have now passed through a general outline of the great science of mind; that immortal, godlike thing, which constitutes the essence of our being. We have but merely hinted at the beauties and excellences of this noble science. It opens for our gaze and labours fields of immeasurable width and infinite length. It uncaps mines of wealth, rare, rich, and new, and explains, in the most attractive manner, all that is old in the knowledge of mind which men possessed before the discovery. It has already established a new and demonstrative philosophy of man, and opened a new era in his progress in mental knowledge. To all, this science is destined to be of great value ; but to the scholar, the teacher, the moralist, and the divine, it is of invaluable worth. It is to overthrow the mouldy structures of time-crusted errors, and rear in their stead a philosophy stirring as life, beautiful as heaven, and solid as the throne of the Eternal.

To you, young men and women, who have listened to me so attentively through this long course of lectures, I bid a hearty God-speed in your seach for knowledge, and struggles for perfection. Go on nobly, bravely on. The gem-paved walks of truth are before you. The bow of eternal promise spans your sky. The fadeless sun of righteousness illumes your pathway. Immortal powers are within you, and imperishable honours are waiting in unseen hands to crown you. Your fathers and mothers are praying for your progress; the world is kneeling in tears and hope for a blessing at your hands. Angels are beckoning you upward, and God is perpetually summoning you to duty.

With such voices, pleading in the majesty and sweetness of

deathless love and hope, can you fail to press onward ? Apply
the principles of our great science to your hearts and lives ;
add to them the sanctions of divine Revelation, and the light of
revealed truth and hope, relying upon your Father's provident
care and impartial love, and you will achieve a crown of light
and blessedness for yourselves, and with loving and devout
hands lay the matchless offering of pure, truthful, and benevolent lives upon the altar of the world.

SUPPLEMENTARY CHAPTER.

By J. BURNS.

Since the first publication of this work, several additions have
been made to the number of phrenological organs, as indicated
in the symbolical head which is placed as a frontispiece to this
edition. It will be observed that the organs located in the
upper part of the forehead are not described in the preceding
lectures. They were introduced by the Fowlers, those most
eminent of practical phrenologists, and it is their system which
is exemplified in the illustration at the beginning of this work,
to which we have referred.

The organ immediately in the centre of the upper part of the
forehead, between Benevolence on the top, and Comparison in
front, has been named Human Nature. Like all other phrenological organs, it is extremely difficult to give it a name
expressive of its true definition under all circumstances,
because the action and quality of every organ is modified in
manifestation by all the other organic conditions. And,
again, it is extremely difficult to give an absolute designation
to that which we are supposed to regard in a relative sense
when it performs a metaphysical function, figurative, so to
speak, of the physical object or condition whose name it bears.
This organ, like all others, is much modified by temperament
and type of head. It may be regarded as a perceptive organ,

operating on the subjective plane somewhat in the same manner as Individuality does in regard to objects. But being purely a metaphysical faculty, it is very much more influenced by temperament than are the perceptives of physical objects. Generally speaking, it may be understood to probe the interior qualities of man, and the probable outcome of events, by its ability to perceive those mental or essential peculiarities of which the external organism and other signs are remote indications. It enables the merchant to adapt goods to his customers ; to read the characteristics of those with whom he has to deal. It enables the employer to appoint suitable men to execute certain kinds of work. It gives the power generally to read mind, and to pry into the thoughts and motives of others. With a low type of brain it tends to an inquisitive, intriguing disposition, which intrudes itself where it has no business, and employs itself with those things that had better be left to others. Associated with the intuitional and sensitive temperaments, it opens the mind to many impressions of what may be called a purely trans-corporeal description. Thus, it anticipates the arrival of visitors, or the fate of friends at a distance. It is prophetical, and its first impressions are sure guides in all matters. It delights to look into man's future, and explore his mental and spiritual destiny. Combined with the critical faculty, it gives fine metaphysical acuteness, and a power of definition and distinction of subtle differences which very much puzzle ordinary minds. Allied to those conditions which incline to the study of Biology and Mental Sciences, it gives great power of diagnosis to the physician and observer of mental phenomena. With such minds the study of human nature becomes a source of real pleasure, and they pursue it successfully because of the intuitive ability to understand man as a scientific fact. Associated with the philanthropic influences of Benevolence, and a genial diffusive temperament, with full development of the affections, this organ gives an intense love for humanity, and the tendency to sympathise with and benefit it in all its forms. Those who have power to win their way and control circumstances are well endowed with this organ.

It enables them, with the aid of the organ on each side of it, to say the right thing at the right time, to adapt the act to the occasion, and perceive the essential point of every case as it comes along. The fineness of this faculty is greatly enhanced by temperament. Thus, one type of organisation will, through this organ, be able to adapt itself more successfully to those that are similar to it, and therefore sympathetic with it, than with those who are out of its organic range, as it were. The organ is most successful in its exercise in those of universal characteristics, enabling them to adapt themselves to almost every species of individual and circumstance. The organ is also modified in its action by the direction of its development. When it is more particularly apparent at that point which unites it to Comparison, it exhibits more of the intellectual and metaphysical character. The middle form of development relates it to the purely intuitional range of function, or the simple perception of those interior conditions to which the organ primarily refers. When the organ of Human Nature inclines towards that of Benevolence, its action leads more in the direction of the philanthropist. The full development of the organ in all its parts presents the most universal form of its action.

It is almost needless to occupy space with the consideration of the facts which arise from its non-development. The person lacks the power of foreseeing much that is immediately related to him, and fails to observe the characteristics of intimate friends, or to adapt his conduct to the circumstances which surround him,—to use a common phrase, he is always "putting his foot into it."

The other organ in the upper part of the forehead to which we refer, is called Agreeableness or Suavity. When largely developed, it gives a height and squareness to the upper part of the forehead. Superficial observers will fail to perceive that they may be misled in the diagnosis of these organs. The square head is really a cranial type known to Dr. Barnard Davis and other anthropologists as the *brachycephalic*, and is what may be called the masculine type of head, giving breadth and grasp to the character, reasoning power, and materialistic ideas based upon actual human experience. The narrow, long type of head, on the contrary, called by these craniologists *dolichocephalic*, is feminine in character, and penetrative, intuitional, and spiritual in tendency, and in thought depends, not so much on theories and rational deductions, as upon impressional and intuitive perceptions which it is able to gather from time to time. Apart from the consideration of these cranial types, it is, however, well to observe that the side organs of the brain are complementary or accessory to those of the centre of the head, and thus while Human Nature perceives the true state of affairs, Agreeableness gives the desire and ability to act in accordance therewith. It may be regarded as a social intuition, enabling man to imitate and adapt himself to those personal peculiarities which he meets with in social intercourse. A person who possesses this development is capable of providing social entertainment when in company, and of supplying small talk when there may be no particular subject on hand. The possessor is playful, youthful, and cheerful in manner. The mind inclines to recreative pastimes.

The examiner is sometimes misled by this form of head in various ways. In the first place, he may find this type associated with small affections, and the subject may be regarded as not of a social character, whereas he may be the soul of social entertainment; but he attends these gatherings, not for the purpose of gratifying his affections, but to exercise those semi-intellectual faculties which enable him to minister to the enjoyment of others and participate in their abilities in return. Such a person is not a domestic stay-at-home individual, nor is he particularly attached to friends, but he goes abroad and into company more professionally, as it were, than from any clinging which he has to individuals. This organ is sometimes mistaken for a development of Causality. The lofty, square, spacious brow may be developed on the range which covers the ground occupied by Agreeableness ; and though the solid intellectual attainments may be but small, with large Language and Ingenuity, this organ can make a little knowledge go a long way, and superficial attainments are made to wear a very prepossessing appearance. Its combinations with Ideality, Wit, Imitation, &c., modify its action, rendering it playful, refined, or dramatic, as the case may be. Devoid of the aid afforded by Human Nature, its playfulness will oftentimes be misplaced. It will romp and disport itself at the wrong time. It is an organ which should be carefully cultivated, especially by those who have to sustain severe mental duties. It relieves the mind from continued strain, and prevents a person becoming prematurely old and weighed down by the burden of life.

There are some other of the organs which are not dwelt upon particularly in the foregoing lectures, but their definition is involved in the consideration of organs with which they generally act in combination.

A spare page at the beginning of the work has been occupied with a new classification of the temperaments, and figures have been introduced to indicate the page on which the relative organs are described in this work. The harmony between the diagram and the lectures is in some instances not very clear, but it is hoped that the new classification will be suggestive to the mind of the studious reader, and lead to further investigation in this important direction. The writer is at present holding classes and making observations, which are discussed in his monthly publication *Human Nature*, and his weekly journal, *The Medium*. Those readers who may be interested in pursuing the subject further, are referred to these publications.

PROF. R. B. D. WELLS

Respectfully announces to the public that he has secured large and commodious premises in the principal thoroughfare of Scarborough, to which he has removed.

One portion of this establishment has been fitted up as a PHRENOLOGICAL MUSEUM, which contains numerous Physiological Diagrams, Skulls, Skeletons, Models, Busts, Casts, Paintings, &c., &c. Public Lectures are delivered at these Rooms several days each week during the Summer months by PROF. and MRS. WELLS, who also give Private Phrenological Examinations, Advice on Health, Choice of Pursuits, the Right Management of Children, Self-Improvement, &c.

The other part of the establishment has been fitted up as a HYGIEAN HOME for the Hygienic Treatment of Invalids, and for the accommodation of visitors who are desirous of securing home comforts, combined with a healthy diet and other conditions which are conducive to health, while in this salubrious town. Tobacco-smoking and intoxication are strictly prohibited, but in other respects the patients may enjoy themselves without restraint, so long as they act with decorum. The treatment of patients is given or personally superintended by MR. and MRS. WELLS, and is not left in the hands of inexperienced attendants, as is the case in many Hygienic Homes. They undertake all curable diseases, but do not take patients into the Home whose diseases are contagious, or that would be likely to affect the other inmates.

During the Winter, PROF. and MRS. WELLS will deliver Popular Lectures on Mental Science and on Health in large towns and cities in the United Kingdom. Halls are engaged each September for the ensuing Winter, and invitations to Lecture from Societies should be sent in on or before the 20th of that month. Persons who wish to ascertain where Mr. WELLS will be Lecturing from October to May each year may do so by sending a stamped and directed envelope to him at Scarbro', on or about the 1st Oct., when a Card of Engagements will be sent per return of post.

Mr. WELLS's Publications may be obtained from JAMES BURNS, No. 15, Southampton Row, London; JOHN and ABEL HEYWOOD, Booksellers, Manchester; and from the Author, Pavilion Place, Westborough, Scarborough.

STANDARD BOOKS AND OTHER APPLIANCES
ON SALE AT
PROF. WELLS' PHRENOLOGICAL & HYGIENIC ESTABLISHMENT,
PAVILION PLACE, SCARBOROUGH,
And also at the close of the Lectures, and during the Day, in Prof. Wells' Consultation Rooms.

1—The SYMBOLICAL HEAD and PHRENOLOGICAL CHART.—
This illustrated map of the head shows at a glance the nomenclature of Combe, as also that of the most eminent living phrenologists. The numbers on the small heads indicate the location of the organs as taught by Gall, Spurzheim, and Combe. The large coloured head symbolizes all the known organs in crania, including all the newly-discovered faculties. This is the most complete symbolical chart published, and is suited for framing. By R. B. D. WELLS. Price 6d.

2—The PHRENOLOGICAL and PHYSIOLOGICAL REGISTER.—
This is a convenient book for practical phrenologists in which to mark
developments. It also contains dietary, marriage, and bath tables
which, when marked either by the physician, doctor, or hydropath, are
invaluable to all classes of the community. By R. B. D. WELLS. Re-
vised and enlarged. 6d.

Also, in course of preparation,

3—A New ILLUSTRATED HANDBOOK of Phrenology, Physiology,
and Physiognomy. By R. B. D. WELLS. Will be ready about January.

4—VITAL FORCE: or, The Evils and Remedies of Perverted Sensuality.
Showing how the health, strength, energy, and beauty of human beings
are wasted, and how preserved. By R. B. WELLS. 1s.

Will shortly be published,

5—MARRIAGE PHYSIOLOGICALLY CONSIDERED, forming Pt. II.
of VITAL FORCE. This book will deal with much that is omitted by
medical writers in general, and will strike at the root of many evils
which manifest themselves in the domestic circle. Suitable for those
about to get married, and also for married persons of both sexes. 1s.

6—PARENTAGE; or, The Population Question Physiologically Considered, forming Part III. of VITAL FORCE. 1s.

7—HEALTH, and HOW to SECURE IT. By R. B. D. WELLS. 1s.

8—HUMAN PHYSIOLOGY: The Basis of Sanitary and Social Science.
By Dr. NICHOLS. 5s. 6d. This is a good book.

9—A WOMAN'S WORK IN WATER CURE. By Mrs. NICHOLS.
1s.; cloth gilt, 2s. 6d.

10—HOW TO COOK. By Dr. NICHOLS. 1s.; paper cover, 6d.

11—HOW TO BEHAVE. By Dr. NICHOLS. 2s. 6d.

12—A SCAMPER ACROSS EUROPE. By Dr. NICHOLS. 6d.

13—THE MYSTERIES of MAN; or, ESOTERIC ANTHROPOLOGY.
This work is written by Dr. NICHOLS (a very able man), and considering its size and price, it is the best book with which I am acquainted,
both as a medical adviser and as a confidential treatise between
physician and patient. All persons who wish to become acquainted
with themselves should secure a copy. 5s.

Chap. 1.—Of Man and his Relations	Chap. 15.—Curative Agencies
,, 2.—The Chemistry of Man	,, 16.—Process of Water Cure
,, 3.—Human Physiology	,, 17.—How to Cure Diseases
,, 4.—Principles of Physiology	,, 18.—Inflammation & Brain Diseases
,, 5.—Of the Organic System	,, 19.—Diseases of the Organs of Res-
,, 6.—The Animal System	piration
,, 7.—The Function of Generation	,, 20.—Diseases of the Organs of Di-
,, 8.—Impregnation	gestion
,, 9.—Morals of the Sexual Relation	,, 21.—Diseases of the Generative
,, 10.—Evolution of the Fœtus	System
,, 11.—Of Pregnancy	,, 22.—Gestation and Parturition
,, 12.—Symptoms of Health	,, 23.—Lactation and the Manage-
,, 13.—The Conditions of Health	ment of Infants
,, 14.—The Cause of Disease	,, 24.—Death and Immortality

14—HOW to LIVE on SIXPENCE A DAY. By Dr. NICHOLS. 6d.

15—HOW COUNT RUMFORD BANISHED BEGGARY FROM BA-
VARIA. By Dr. NICHOLS. 6d.

16—FORTY YEARS of AMERICAN LIFE. By Dr. NICHOLS. 5s. 6d.

17—The NEW ILLUSTRATED SELF-INSTRUCTOR in PHRENO-
LOGY and PHYSIOLOGY. By O. S. and L. N. FOWLER. 2s.

3

67—PHRENOLOGICAL BUSTS, in China, 10s. 6d. each; in Plaster of Paris, from 1s. to 6s.

68—SITZ BATHS. Unpainted, 16s.; painted, 18s. and 20s.

69—BATH SPONGES, and all kinds of Water Cure Appliances, are supplied at reasonable prices.

70—Professor R. B. D. WELLS, or his assistants, supply the following Medical Instruments at the close of each Lecture, or in the Ante-Room during each day:—

71.—ENEMAS or SYRINGES, with Instructions for use. The following are the Prices of Enemas supplied by Mr. WELLS:

72—No. 1. The Patent Disconnecting, Enema, with Pipe for the Anus or Seat, and Vagina or Womb. 8s. 6d.

73—No. 2. Patent Disconnecting Enema, with Pipe for Anus only, 7s. 6d.

74—No. 3½. English make (not disconnecting), with Pipe for Anus, 6s. 6d.

75—No. 3. English make (not disconnecting), with Pipe for Anus, 5s. 6d.

76—No. 4. English make (not disconnecting), with Pipe for Anus, 4s. 6d.

77—No. 5. English make (bottle shape), with Pipe for Anus, 3s. 6d.

There is scarcely any case of disease in which Injections, once or twice a day, may not be used to advantage.

When the bowels are in a very torpid condition, these Injections may be taken frequently, and to the extent of two or three pints; but in most cases one pint is sufficient to cause the bowels to act freely.

All persons liable to sickness should have a good syringe for their own use; but there are many Syringes or Enemas made that are worse than useless, inasmuch as the valves soon become impaired, and the result is a griping pain in the bowels, caused by the injecting of air instead of water, hence everyone should be particular in the selection of a suitable instrument.

The Patent Disconnecting Enemas are the only ones that Mr. Wells can recommend, inasmuch as they are well made, and if the valves corrode by use they can be disconnected, the valves liberated, and set to work in perfect condition in three minutes—a very easy process, and learned in one minute. This Enema is not likely to get out of order as readily as most other makes, and it generally lasts many years if carefully used If the Enema does not work freely, unscrew the tube at the end of the ball, and liberate the small leaden valve at each end of it, and be sure to put it together again in the same way as before.

78—CHEST EXPANDERS, suitable for persons with Weak Chests. From 1s. 6d. to 6s. each.

79—ELASTIC STOCKINGS, &c.

80—SPIRIT STOVE and KETTLE, for Boiling a Pint of Water in five minutes without making a fire in the sick-room. 2s. 6d.

81—STOMACH CANS, 3s 6d. LEG BATHS, 7s. 6d.

82—Best SWANSDOWN CALICO, for Compresses, 1s. per yard. 1¼ yards are required for each Compress.

83—BANDAGES, 1s., 1s. 6d., and 2s. each, according to length.

84—GAS STOVES, for Hot-Air Baths, 5s.

85—URINALS, suitable for persons who are unable to retain their water.

86—DOVER EGG BEATERS, for Beating up Eggs or Mixing Butter, so as to secure light and digestible food. 2s.

8—ABDOMINAL SUPPORTS, MACINTOSH BANDAGES, &c., &c.

Phrenological and Physiological Works got to order at current prices.

A Catalogue of Standard Works (for Sale) Free on application.

THE SCIENCE OF A NEW LIFE.

BY JOHN COWAN, M.D.

Printed from beautifully clear, new type, on fine calendered, tinted paper, in 1 vol. of over 400 pages, 8vo, containing 100 first-class engravings, and a fine steel-engraved frontispiece of the author. Bound in cloth, bevelled boards, gilt back and side stamp. Price 12s. 6d.

TABLE OF CONTENTS.

CHAPTER I. Marriage and its advantages.—II. Age at which to marry.—III. The law of choice.—IV. Love analysed.—V. Qualities the man should avoid in choosing.—VI. Qualities the woman should avoid in choosing.—VII. The anatomy and physiology of generation in woman.—VIII. The anatomy and physiology of generation in man.—IX. Amativeness: its use and abuse.—X. The prevention of conception.—XI. The law of continence.—XII. Children, their desirability.—XIII. The law of genius.

PART II.—THE CONSUMMATION.

CHAPTER XIV. The conception of a New Life.—XV. The physiology of inter-uterine growth.—XVI. Period of gestative influence.—XVII. Pregnancy, its signs and duration.—XVIII. Disorders of pregnancy.—XIX. Confinement.—XX. Management of mother and child after delivery.—XXI. Period of nursing influence.

PART III.—WRONGS RIGHTED.

CHAPTER XXII.—Fœticide.—XXIII. Diseases peculiar to women.—XXIV. Diseases peculiar to men.—XXV. Masturbation.—XXVI. Sterility and Impotence.—XXVII. Subjects of which more might be said.—XXVIII. A happy married life, how secured.

PERSONAL AND NEWSPAPER NOTICES.

[*From the " Christian Union,"* HENRY WARD BEECHER, *Editor.*]

A new edition of "The Science of a New Life" gives us the opportunity of saying that it seems to us to be one of the wisest and purest and most helpful of those books which have been written in recent years with the intention of teaching men and women the truths about their bodies, which are of peculiar importance to the morals of society. It will be understood that we here refer to treatises on sexual physiology. No one can begin to imagine the misery that has come upon the human family solely through ignorance upon this subject. Of course, only a man who is more than learned, who is wise and good also, can safely be trusted with the duty of writing such a book. The spirit in which Dr. Cowan has written is apparently that of earnest devotion to the welfare of mankind.

[*Extract of a letter from* ROBERT DALE OWEN *to the Author.*]

I thank you much for the brave book you were so kind as to send me. The subjects upon which it touches are among the most important of any connected with Social Science, and the world is your debtor for the bold stand you have taken. Yours sincerely, ROBERT DALE OWEN.

[*From* JUDGE J. W. EDMONDS, *ex-Chief Justice of the Supreme Court, New York.*]

I have read the work "The Science of a New Life," by Dr. John Cowan, and I ought not to withhold from you the expression of my approbation of it. I would have given a good deal for the knowledge it contains in my boy days—some sixty years ago, and I rejoice greatly that it has at length been put in a form accessible to all.   J. W. EDMONDS.

[*From the "Banner of Light," Boston.*]

We welcome a publication of this sort with undisguised sincerity, thank- ' ful that the time at last has come when fundamental and radical physiological truths may be told to the people plainly. Had such books been placed in the hands of younger men two or three generations ago, their effect would have been visible enough in the physical character and habits of the men of to-day.

Noted Men and Women who have endorsed the principles it contains.

Please remember that these signatures, as well, in fact, as all the endorsements the book has received, have *not* been secured through favouritism or personal friendship, but entirely and altogether on the merits of the book

Rev. E. H. Chapin, D.D., Editor " Christian Leader," New York.
Grace Greenwood, Author and Lecturer.
Francis E. Abbott, Editor "Index," Boston, Mass.
Rev. Wm. R. Alger, Author and Lecturer, Boston, Mass.
Rev. W. T. Clark, "The Daily Graphic," New York.
Rev. Charles F. Deems, D.D., Editor "Christian Age."
Judge J. W. Edmonds, Ex-Chief Justice of the Supreme Court, N. Y.
Rev. George H. Hepworth, Church of the Disciples, New York.
Oliver Johnson, Managing Editor " Christian Union," New York.
Dr. Dio Lewis, Author and Leturer, Boston, Mass.
Mrs. Clemence S. Lozier, M.D., Dean of the New York Medical College for Women.
Gerald Massey, Poet and Lecturer, England.
Theodore Tilton, Editor " Golden Age."
Moses Coit Tyler, Literary Editor " Christian Union."
Mrs. Caroline M. Severance, West Newton, Mass.
Mrs. Elizabeth Cady Stanton, New York.

Rev. O. B. Frothingham, New York.	Wendell Phillips, Boston, Mass.
Mrs. Frances Dana Gage, New York.	Parker Pillsbury.
Wm. Lloyd Garrison, Boston, Mass.	Hon. Gerrit Smith.
Hon. Robert Dale Owen.	Dr. H. B. Storer, Boston Mass.
James Parton, New York.	Rev. Warren H. Cudworth, Boston.

Now ready, in neat cloth, eighty pages, price 1s.

HEALTH HINTS;
SHOWING HOW TO ACQUIRE AND RETAIN BODILY SYMMETRY, HEALTH, VIGOR, AND BEAUTY.

TABLE OF CONTENTS:

. HEALTH, WEALTH, AND HAPPINESS
SERIES OF POPULAR HANDBOOKS
FOR TOWN AND COUNTRY.

Third Edition, 93 *pp.,* price 6*d.*

ILLNESS : ITS CAUSE AND CURE,

Showing how to preserve health and cure diseases by a safe, scientific, pleasant, and efficient means within the reach of all.

How to Preserve Health is a matter of no small importance, nor is it an Utopian undertaking. Nearly all diseases are preventible, and the fraction of time and money spent in acquiring the necessary knowledge is insignificant compared with the loss and suffering incurred by ill-health, doctors, and drugs.

How to Cure Diseases normally is indicated by the means required to preserve health. Such modes of cure are :—

Safe.—Being in accordance with the laws of health, they cannot possibly destroy the patient, or undermine the constitution, as the common practice of administering poison does.

Scientific.—The remedies propounded in this book are based upon the *nature of disease,* and the demands of the system in respect to regaining the normal condition. Hence, dangerous courses of experiments are superseded by a certain means producing the desired result. This practical knowledge will prove the death-blow to all kinds of medical quackery and malpractice.

Pleasant are such means, and grateful to the diseased condition as food is to the hungry, drink to the thirsty, or rest to the weary. No disgusting draughts, painful operations, or enfeebling processes, but the whole is regenerating and restorative.

Efficient in all cases where cure is possible, is this system. Under it acute diseases, small-pox, fevers, diphtheria, bronchitis, rheumatism, &c., and all common ailments lose their virulent character ; and by observing the rules of health laid down, they might be banished from the land, and with them the dreaded cholera.

These means are within the reach of all. The poorest in the land may understand the system and avail themselves of its blessings. Sanitary associations should be formed in each town, and missionaries employed to teach it to those who cannot read and investigate these simple phenomena for themselves.

Send 7 stamps for a sample copy at once, while you are well, and do all you can to spread it amongst your friends. They are sold at a reduced price in quantities for distribution.

By the author of " Illness : its Cause and Cure."

SIMPLE QUESTIONS & SANITARY FACTS,
For the Use of the People.
Price One Shilling.

This work, in the form of questions and answers, in a very pleasing manner conveys a vast amount of information on various branches of physical science, health, dietetics, morals, and remedial agencies.

WORKS BY R. T. TRALL, M.D.

SEXUAL PHYSIOLOGY:

A SCIENTIFIC AND POPULAR EXPOSITION OF THE FUNDAMENTAL PRINCIPLES OF SOCIOLOGY.

It contains 16 chapters, extends over 312 pages, illustrated by 80 engravings; gives the complete anatomy and physiology of the sexual organs, origin of life, and everything connected with impregnation and generation, according to the latest discoveries in science; strongly bound in cloth.

CONTENTS.

Embriology	Regulation of number of Offspring
The Organs of Plants	Theories of Population
Physiology and Menstruation	Hereditary Transmission
Conception	Rights of Offspring
Indications of Pregnancy	Beautiful Children
Development of the Germ	Good Children
Circulation of the Fœtus	Intermarriages
The Science of Propagation	Temperament
Rationale of Labour	Miscegenation
Lactation and Pregnancy	Physiology of Marriage
Law of Sex	Courtship
Theory of Sex	Union for Life
Experiments	Choosing a Husband or Wife
Latest Theories	Marriageable Age, &c., &c.

These topics are treated from a scientific and elevated standpoint. The position of the author in the medical world (not being a drug doctor) precludes the chance of quackery. There are no disgusting details of diseases nor immoral associations, but just such useful, aye, and indispensable information as would enable many well-meaning men and women to afford to each other and to posterity the obligations which the closest and most endearing associations demand. There is no relationship so important as that which underlies the formation of a new being, and there is none in which there is so much popular ignorance and abuse. This, then, is a work of the greatest importance, and should be widely diffused by all lovers of morality and human progress. Price 5s., or six copies for the price of five.

HYDROPATHIC ENCYCLOPCEDIA; the most complete work on this subject; with 300 engravings. Price 16s.

HYGIENIC HANDBOOK. Practical Guide for the Sick-room. 8s. 6d.

DIGESTION AND DYSPEPSIA. A Complete Explanation of the Physiology of the Digestive Processes, Symptoms, and Treatment of Dyspepsia and other Disorders of the Digestive Organs; engravings. 4s.

FAMILY GYMNASIUM; Gymnastic, Calisthenic, Kinesipathic, and Vocal Exercises. Many illustrations. 7s. 6d.

HYDROPATHIC COOK-BOOK on Hygienic Principles. 6s.

THE BATH: its History and Uses in Health and Disease. 1s. 6d.

WATER CURE FOR THE MILLION; the Process Explained. 1s. 6d.

TRUE HEALING ART, or Hygienic versus Drug Medication. 1s. 6d.

THE TRUE TEMPERANCE PLATFORM. Medicinal Use of Alcohol discussed. 1s.　　THE ALCOHOLIC CONTROVERSY. 2s.

DISEASES OF THE THROAT AND LUNGS, and their Treatment. 1s.

· EUROPEAN AGENCY FOR THE STANDARD WORKS
OF
SAMUEL R. WELLS AND CO.,
389, Broadway, New York.

AMERICAN PHRENOLOGICAL JOURNAL and LIFE ILLUS-
TRATED, with which is incorporated the "Science of Health."
Devoted to Ethnology, Physiology, Phrenology, Physiognomy, Psychology, Sociology, Hygiene, Biography, Education, Art, Literature.
1s. 6d; post-free, 16s. per annum.

DEFENCE of PHRENOLOGY. By Boardman. 6s.

EDUCATION and SELF-IMPROVEMENT COMPLETE; comprising
"Physiology—Animal and Mental," "Self-Culture and Perfection of
Character," "Memory and Intellectual Improvement." 16s.

PHRENOLOGY PROVED, ILLUSTRATED, and APPLIED. Embra-
cing an Analysis of the Primary Mental Powers in their Various
Degrees of Development, and Location of the Phrenological Organs.
Amply Illustrated. 492 pp. 7s. 6d.

NEW PHYSIOGNOMY; or, Signs of Character, as manifested through
Temperament and External Forms, and especially in the "Human
Face Divine," With more than One Thousand Illustrations. 21s.

HOW to READ CHARACTER. A New Illustrated Hand-book of
Phrenology and Physiognomy, with upwards of 170 Engravings. 191 pp.
Paper, 4s. Cloth, 5s.

CHART of PHYSIOGNOMY, Illustrated. Designed for Framing. 1s.

PHRENOLOGY and the SCRIPTURES. By Rev. J. Pierpont. 1s.

LECTURES on PHRENOLOGY. By George Combe. With Notes. An
Essay by Boardman. 7s. 6d.

ANNUALS of PHRENOLOGY and PHYSIOGNOMY for Nine Years.
By S. R. Wells, Editor of the *Phrenological Journal*. 8s. 6d.

WEDLOCK, or the Right Relations of the Sexes. 6s.

ACCIDENTS and EMERGENCIES. Guide for Treatment of Wounds. 1s.

CHILDREN : their Management in Health and Disease. Descriptive and
Practical. Shew. 7s. 6d. ·

FRUITS and FARINACEA, the PROPER FOOD of MAN. With Notes
by Trall. 7s. 6d.

FAMILY PHYSICIAN. A Ready Prescriber and Hygienic Adviser. By
Joel Shew, M.D. 16s.

MEDICAL ELECTRICITY. A Manual showing the most scientific and
rational application to all forms of Disease of the different combinations of Electricity, Galvanism, Electro-Magnetism, Magneto-Electricity
and Human Magnetism. By White. 9s.

MIDWIFERY and the DISEASES of WOMEN. With General Management of Childbirth, Nursery, &c. For Wives and Mothers. 7s. 6d.

AN EXPOSITION of the SWEDISH MOVEMENT-CURE. By Dr.
Taylor. 7s. 6d.

MOTHER'S HYGIENIC HAND-BOOK, for the Normal Development and Training of Women and Children, and the Treatment of their Diseases with Hygienic Agencies. By R. T. TRALL, M.D. 5s.

NOTES on BEAUTY, VIGOUR and DEVELOPMENT; or, How to Acquire Plumpness, Strength, and Beauty of Complexion. 6d.

PRACTICE of the WATER-CURE. A Detailed Account of the various processes. 2s.

THE SCIENCE of HUMAN LIFE. By SYLVESTER GRAHAM. With a Copious Index and Biographical Sketch of the Author. 14s.

SOBER and TEMPERATE LIFE. Discourses and Letters of CORNARO. With a Biography. He died at 100 years of age. 2s.

THE HUMAN FEET. Their Shape, Dress, and Care. Of universal interest. 5s.

THE PARENTS' GUIDE; or, Human Development through Inherited Tendencies. By Mrs. HESTER PENDLETON. 6s.

THE INDISPENSABLE HANDBOOK for HOME IMPROVEMENT. Comprising "How to Write," "How to Talk," "How to Behave," and "How to do Business." Excellent. 10s.

WEAVER'S WORKS. Comprising "Hopes and Helps," "Aims and Aids," "Ways of Life." By Rev. G. S. WEAVER. One vol., 12s.

The following three volumes may be had separately:—

HOPES and HELPS for the YOUNG of BOTH SEXES. 6s.

AIMS and AIDS for GIRLS and YOUNG WOMEN, on the Duties of Life. 6s.

WAYS of LIFE, showing the Right Way and the Wrong Way. 4s.

LIFE AT HOME; or, the Family and its Members, Husbands, Wives, Parents, Children, Brothers, Sisters, Employers and Employed. By Rev. WM. AIKMAN, D.D. 6s.

ORATORY—SACRED and SECULAR: the Extemporaneous Speaker. 6s.

LIBRARY of MESMERISM and PSYCHOLOGY. Comprising the Philosophy of Mesmerism, Clairvoyance, Mental Electricity.—Fascination, or the Power of Charming.—The Macrocosm, or the Universe Without.—The Philosophy of Electrical Psychology; Doctrine of Impressions; Mind and Matter. Psychology, or the Science of the Soul, considered Physiologically and Philosophically. Large vol., 16s.

FOOTPRINTS of LIFE; or, Faith and Nature Reconciled. 5s.

A SELF-MADE WOMAN; or, Mary Idyl's Trials and Triumphs. By E. M. BUCKINGHAM. 6s.

THE NEW ILLUSTRATED SELF-INSTRUCTOR in Phrenology Physiology, and Physiognomy. With 100 Portraits. 175 pages. Cloth, 2s.

LECTURES ON MAN: as explained by Phrenology, Physiology, Physiognomy and Ethnology. By L. N. FOWLER, 21 in Number, 2d. each, or in one volume, 5s.

WOMAN: Her Destiny and Material Relations. By Mrs. L. F. FOWLER, 6d.

MR. FOWLER'S New and Improved PHRENOLOGICAL BUST. With upwards of 100 divisions. In China, 10s. 6d.: Plaster, 6s. Other Busts at various prices, down to 1s. each.

RECENT WORKS ON SPIRITUALISM.

MIRACLES AND MODERN SPIRITUALISM. By ALFRED R. WALLACE, F.R.G.S., F.Z.S., Author of "Travels on the Amazon and Rio Negro," "Palm Trees of the Amazon," "The Malay Archipelago," &c., &c. Cloth, 5s.; handsomely gilt, 7s. 6d. Embracing:

> I.—"AN ANSWER TO THE ARGUMENTS OF HUME, LECKY, AND OTHERS AGAINST MIRACLES."
>
> II.—"THE SCIENTIFIC ASPECTS OF THE SUPERNATURAL," much enlarged, and with an Appendix of Personal Evidence.
>
> III.—"A DEFENCE OF MODERN SPIRITUALISM," reprinted from the *Fortnightly Review*.

RESEARCHES in the PHENOMENA of SPIRITUALISM. By WILLIAM CROOKES, F.R.S., &c. 16 illustrations. Cloth, 5s.; or in 3 parts, 1s. each.

> I.—SPIRITUALISM VIEWED BY THE LIGHT OF MODERN SCIENCE, and EXPERIMENTAL INVESTIGATIONS IN PSYCHIC FORCE.
>
> II.—PSYCHIC FORCE AND MODERN SPIRITUALISM: a Reply to the *Quarterly Review* and other critics.
>
> III.—NOTES ON AN INQUIRY INTO THE PHENOMENA CALLED SPIRITUAL DURING THE YEARS 1870-73.

ARCANA OF SPIRITUALISM: A Manual of Spiritual Science and Philosophy. By HUDSON TUTTLE. New Edition. 5s.

RESEARCHES IN MODERN SPIRITUALISM. By M.A. (Oxon), appearing monthly in *Human Nature*, a journal of Zoistic Science and Popular Anthropology. 6d.

ORATIONS through the Mediumship of Mrs. CORA L. V. TAPPAN; The New Science—Spiritual Ethics—containing upwards of 50 Orations and Poems. 720 pages, Full gilt, with photograph, 10s. 6d.; handsome cloth, 7s. 6d.

LETTERS AND TRACTS ON SPIRITUALISM. By JUDGE EDMONDS. Memorial Edition, with Memoir and Passing Away of the Author; and Discourses by Theodore Parker and Judge Edmonds, through Mrs. TAPPAN. Cloth, 3s. 6d.

EXPERIENCES IN SPIRITUALISM: Records of Extraordinary Phenomena through the most Powerful Mediums, with Photograph of the Author. By CATHERINE BERRY. 3s. 6d.

HAFED, PRINCE OF PERSIA: HIS EARTH-LIFE AND SPIRIT-LIFE. Trance Communications through D. DUGUID, by a Spirit who was a personal follower of Jesus. Illustrated with lithographs of Direct Spirit-Drawings and many examples of Direct Writing, 10s.

WHERE ARE THE DEAD? or Spiritualism Explained. By F. A. BINNEY. 3s.

REPORT ON SPIRITUALISM OF THE COMMITTEE OF THE LONDON DIALECTICAL SOCIETY. 5s.

THE SEERS OF THE AGES, or Spiritualism Past and Present. By J. M. PEEBLES. 5s.

SCIENTIFIC MATERIALISM EXAMINED and REFUTED.
Being a Reply to the Address of Professor Tyndall delivered
before the British Association in August, 1874. at Belfast. By
GEORGE SEXTON, LL.D. Price 1s.; cloth, 2s. 6d.

STARTLING FACTS IN MODERN SPIRITUALISM. By
N. B. WOLFE, M.D. Illustrated with Portraits on Steel, Spirit-
Writings, Diagrams, &c. 550 pp., toned paper, fine binding 12s.

IDENTITY OF PRIMITIVE CHRISTIANITY & MODERN
SPIRITUALISM. By DR. CROWELL. 2 vols., 10s. each.

INCIDENTS IN MY LIFE. By D. D. HOME. First series,
7s. 6d.; second series, 10s.

CONCERNING SPIRITUALISM. By GERALD MASSEY. 2s.

SPIRITUALISM, its Facts and Phases. By J. H. POWELL. 2s.

SCEPTICISM AND SPIRITUALISM; or the Experiences of
a Sceptic. 1s. 6d.; cloth, 2s. 6d.

OUTLINES of MODERN SPIRITUALISM. By T. P. BARKAS.
1s. 6d.

MODERN AMERICAN SPIRITUALISM : a Twenty Years'
Record of the Communion between Earth and the World of
Spirits. By EMMA HARDINGE. 15s.; cheap edition, 10s.

THE CLOCK STRUCK THREE; showing the Harmony
between Christianity, Science, and Spiritualism. By S. WATSON,
D.D. 6s.

THE COMPLETE WORKS OF A. J. DAVIS, comprising
Twenty-seven Uniform Volumes, all neatly bound in cloth.

> Nature's Divine Revelations, 15s.
> The Physician. Vol. I. Gt. Harmonia, 7s. 6d.
> The Teacher. ,, II.   ,,   7s. 6d.
> The Seer.   ,, III.   ,,   7s. 6d.
> The Reformer. ,, IV.   ,,   7s. 6d.
> The Thinker.   ,, V.   ,,   7s. 6d.
> Magic Staff—An Autobiography of A. J. Davis, 7s. 6d.
> A Stellar Key to the Summer Land, 3s. 6d.
> Arabula, or Divine Guest, 7s. 6d.
> Approaching Crisis, or Truth versus Theology, 5s.
> Answers to ever-recurring Questions from the People, 7s. 6d.
> Children's Progressive Lyceum Manual, 3s.
> Death and the After-Life, 3s. 6d.
> History and Philosophy of Evil, 3s. 6d.
> Harbinger of Health, 7s. 6d.
> Harmonial Man, or Thoughts for the Age, 3s. 6d.
> Events in the Life of a Seer. (Memoranda). 7s. 6d.
> Philosophy of Special Providences, 2s. 6d.
> Free Thoughts concerning Religion, 3s. 6d.
> Penetralia, containing Harmonial Answers, 7s. 6d.
> Philosophy of Spiritual Intercourse. 6s.
> The Inner Life, or Spirit Mysteries Explained, 7s. 6d.
> The Temple—on Diseases of Brain and Nerves, 7s. 6d.
> The Fountain, with Jets of New Meanings, 5s.
> Tale of a Physician, or Seeds and Fruits of Crime, 5s.
> The Sacred Gospels of Arabula, 5s.
> Diakka, and their Earthly Victims, 2s. 6d.

PEOPLE FROM THE OTHER WORLD. On the Material-
isation of Spirit-Forms, as seen at the Eddy Homestead. By
Col. OLCOTT. Seventy-four illustrations, 500 pages. 12s. 6d.

THE HISTORY of THE SUPERNATURAL in ALL AGES and NATIONS. By W. HOWITT. 2 vols., 18s.

SUPERMUNDANE FACTS IN THE LIFE OF THE Rev. J. B. FERGUSON, D.D. 5s.

PLANCHETTE: or, Despair of Science. By EPES SARGENT. 6s.

THE PROOF PALPABLE OF IMMORTALITY. By EPES SARGENT. 5s.

AROUND THE WORLD: or, Travels in Polynesia, China, India, Arabia, Egypt, Syria, and other "Heathen" Countries. By J. M. PEEBLES, 10s.

JESUS: MYTH, MAN, OR GOD: or the Popular Theology and the Positive Religion Contrasted. By J. M. PEEBLES. 1s. 6d.; cloth, 2s. 6d.

THE CAREER OF RELIGIOUS IDEAS. By H. TUTTLE. 2s. 6d.

BIOGRAPHY of Mrs. J. H. CONANT, the World's Medium of the Nineteenth Century. 7s. 6d.

FLASHES OF LIGHT FROM THE SPIRIT-LAND, through the Mediumship of Mrs. J. H. CONANT. 7s. 6d.

STRANGE VISITORS. A series of Original Papers, embracing Philosophy, Science, Government, Religion, Poetry, Art, Fiction, Satire, Humour, Narrative, and Prophecy. By Eminent Spirits, dictated through a Clairvoyant. 6s.

FOOTFALLS on the BOUNDARY of ANOTHER WORLD. By R. D. OWEN. 7s. 6d.

THE DEBATEABLE LAND BETWEEN THIS WORLD AND THE NEXT. By R. D. OWEN. 7s. 6d.

HINTS FOR THE EVIDENCES OF SPIRITUALISM. By M.P. 2s. 6d.

THE MENDAL; a Mode of Oriental Divination, disclosing remarkable revelations in Biology and Psychology; giving the true key to Spirit-Agency, and the nature of Apparitions, and the connection between Mesmerism and Spiritism. And in Part Second, "Materialism," the Source and Necessary Attendant on Social Disorganisation. By EDWARD B. B. BARKER, Esq., a British Vice-Consul. 7s. 6d.

PSYCHOPATHY, or THE TRUE HEALING ART. By JOSEPH ASHMAN. With photograph of Author, by HUDSON, showing a halo of healing aura over his hands. Second Edit., cloth, 2s. 6d.

WILL-ABILITY; or Mind and its Varied Conditions and Capacities. By JOSEPH HANDS, M.R.C.S. 2s. 6d.

NATURE'S REVELATIONS OF CHARACTER: or, Physiognomy Illustrated. By J. SIMMS, M.D. A large and handsome volume, containing 270 engravings. 21s.

LECTURES ON MENTAL SCIENCE. By G. S. WEAVER. Reprint of a Popular and Eloquent American Treatise on Phrenology. Original Edition, 5s.; New Edition, 2s. 6d.

ETHNOLOGY AND PHRENOLOGY AS AN AID TO THE
HISTORIAN. By J. W. JACKSON. 4s.

MAN: Considered Physically, Morally, Intellectually and Spiritually. By J. W. JACKSON. 5s.

HESPERIA: an Epic of the Past and Future of America. By
Mrs. TAPPAN. 1 vol., cloth, 6s.

WORKS BY THOMAS LAKE HARRIS.
Arcana of Christianity: an Unfolding of the Celestial Sense of the
   Divine Word. Part III.—The Apocalypse, Vol. I.—all published, 6s.
The Breath of God with Man: an Essay on the Grounds and Evidences
   of Universal Religion. 8vo, cloth, 1s. 6d.
The Great Republic: a Poem of the Sun. 8vo, hf. morocco, 6s.
A Lyric of the Morning Land. Cap. 8vo, cloth, 2s.; paper, 1s. 6d.
A Lyric of the Golden Age.   Cloth, 8vo, gilt edges, 5s.
   Ditto   Calf, neat, 7s. 6d.

IPHIGENIA and other Poems. By HENRY PRIDE. 3s.

THE SPIRITUAL HARP and SPIRITUAL LYRE, in 1 vol.
The finest assortment of Spiritual Hymns ever published. 350
pages, 2s. 6d. Morocco, highly gilt and finished, for presents, 5s.

NINE THOUSAND RECOGNISED SPIRIT-PHOTOGRAPHS
given gratis to the readers of *Human Nature.* The Photographic
Series, containing six genuine Spirit-Photographs, with signed testimonies of the sitters and elaborate articles by M.A. (Oxon.), post
free, 2s. 10d.

HUMAN NATURE: A Monthly Record of Zoistic Science;
high-class Magazine for Spiritualists. 6d. monthly; 7s. yearly.

SPIRIT-MEDIUMS AND CONJURERS. An Explanation of
the Tricks of all Conjurers who pretend to Expose Spiritualism:
How to escape from a Corded Box—How to get out of the
Stocks—The Magic Cabinet—How to get out of Sealed and
Knotted Ropes, and perform the Conjurers' so-called "Dark
Seances"—How to Perform the Blood-Writing on the Arm, and
read Names written on Papers by the Audience. The phenomena
attending Spirit-Mediums are clearly defined and shown to be
quite distinct from the tricks of Conjurers. 2d.; post free, 2½d.

Now Publishing, in sixteen parts, 2s. 6d. each,
ANACALYPSIS: an Attempt to Draw Aside the Veil of the Saitic Isis; or,
an Inquiry into the Origin of Languages, Nations, and Religions. By
GODFREY HIGGINS, Esq., F.S.A., F.R.Asiat.Soc., F.R.Ast.S. (Late of Skellow Grange, near Doncaster.) This magnificent work has always been
scarce, but it is now out of print. Copies in two huge volumes have sold
freely at prices ranging from five to fifteen guineas. It is now in course of
publication in sixteen parts, price 2s. 6d. each, or in volumes, price £2 2s.
the set.

RULES FOR THE SPIRIT-CIRCLE. By EMMA HARDINGE. 1d.
THE SPIRIT-CIRCLE AND THE LAWS OF MEDIUMSHIP. By EMMA HARDINGE. 1d.
THE PHILOSOPHY OF DEATH. By A. J. DAVIS. 2d.
MEDIUMS AND MEDIUMSHIP. By T. HAZARD. 2d.
WHAT SPIRITUALISM HAS TAUGHT. By WILLIAM HOWITT. 1d.
CONCERNING THE SPIRIT-WORLD. By J. J. MORSE. 1d.
SPIRITUALISM AS AN AID AND METHOD OF HUMAN PROGRESS. By J. J.
MORSE. 1d.
A SCIENTIFIC VIEW OF MODERN SPIRITUALISM. By T. GRANT. 1d.
WHAT IS DEATH? By JUDGE EDMONDS. 1d.

THEODORE PARKER IN SPIRIT-LIFE. By Dr. WILLIS. 1d.

SPIRIT-MEDIUMS AND CONJURERS. By Dr. SEXTON. 2d.

GOD AND IMMORTALITY VIEWED IN THE LIGHT OF MODERN SPIRITUALISM. By Dr. SEXTON. 6d.

IS SPIRITUALISM THE WORK OF DEMONS? By T. BREVIOR. 2d.

CONCERNING MIRACLES. By T. BREVIOR. 3d.

IMMORTALITY IN HARMONY WITH MAN'S NATURE AND EXPERIENCE: Confessions of Sceptics. By T. BREVIOR. 3d.

THE GOSPEL OF HUMANITY; or, the Connection between Spiritualism and Modern Thought. By GEORGE BARLOW. 6d.

SPIRITUALISM PROVED BY FACTS: Report of a Two Nights' Debate between C. Bradlaugh, Secularist, and J. Burns, Spiritualist. 6d.

SPIRITUALISM, THE BIBLE, AND TABERNACLE PREACHERS. By J. BURNS. A Reply to Dr. Talmage's " Religion of Ghosts." 2d.

THE SYMPATHY OF RELIGIONS. By T. W. HIGGINSON. 2d.

EXPERIENCES OF A SEPTUAGENARIAN. By J. JUDD. 1d.

CLAIRVOYANCE. By PROFESSOR DIDIER. 4d.

DEATH, IN THE LIGHT OF THE HARMONIAL PHILOSOPHY. By MARY F. DAVIS. 1d.

SUGGESTIONS FOR A PUBLIC RELIGIOUS SERVICE IN HARMONY WITH MODERN SCIENCE AND PHILOSOPHY. 6d.

BUDDHISM AND CHRISTIANITY: Remarks on the Opinions of the Right Rev. Bishop Claughton on Buddhism. By a Sceptic. 6d.

THE LANGUAGE OF THE RAINBOW OF COLOUR. By F. WILSON, Arch-keeper of the Cardinal Blue. Beautifully Coloured. 4d.

THE IDEAL ATTAINED: being the Story of Two Steadfast Souls, and How they Won their Happiness and Lost it Not; A tale of pure love and true marriage. By ELIZA W. FARNHAM. 5s.

SOCIAL FETTERS. a Tale of Thrilling Interest and Pure Purpose. By Mrs. EDWIN JAMES. 3s. 6d.

BEAUTIFUL ENGRAVINGS FOR FRAMING.

THE ORPHAN'S RESCUE. 12s. 6d.

LIFE'S MORNING AND EVENING. 12s. 6d.

TEN SPIRITUAL COMMANDMENTS AND LAWS OF RIGHT; with the Creed of the Spirits: A Declaration of Moral and Philosophical Principles given by the Spirits through EMMA HARDINGE. Lithographed in tints, on a beautiful artistic design, illustrating the Law of Inspiration and Spirit-Teaching. Very appropriate for framing and hanging on the wall. 3s. 6d.

DR. DODS' GREAT WORK ON MESMERISM.

Now ready, in handsome illustrated cloth binding, two vols. in one; price 3s. 6d. (The original editions sell for 8s.)

The whole of Dods' celebrated Lectures, embracing Six Lectures on

THE PHILOSOPHY OF MESMERISM,

and Twelve Lectures on the

PHILOSOPHY OF ELECTRICAL PSYCHOLOGY,

including the Lecture, worth all the money, entitled

"THE SECRET REVEALED, SO THAT ALL MAY KNOW HOW TO EXPERIMENT WITHOUT AN INSTRUCTOR."

A TREATISE ON

ALL THE KNOWN USES OF ORGANIC MAGNETISM,

PHENOMENAL AND CURATIVE. By Miss Leigh Hunt. 2d.

WORKS ON VACCINATION.

Have you been Vaccinated, and what Protection is
it against Smallpox? By WILLIAM J. COLLINS, M.D. 6d.

Lecture on Vaccination. A most comprehensive work. By
MISS LEIGH HUNT. 4d.

FOWLER'S WORKS

ON

PHRENOLOGY, PHYSIOLOGY, &c.

Cheap Edition.

AMATIVENESS; or, Evils and Remedies of Excessive and Perverted Sexuality.
With advice to the Married and Single. By O. S. Fowler. 3d.

LOVE AND PARENTAGE; applied to the Improvement of Offspring. Including important directions and suggestions to Lovers and the Married, concerning the strongest ties, and the most sacred and momentous relations of life.
By O. S. Fowler. 3d.

MATRIMONY; or, Phrenology and Physiology applied to the selection of Congenial Companions for life. Including directions to the Married for living
affectionately and happily. By O. S. Fowler. 3d.

PHYSIOLOGY, ANIMAL AND MENTAL; applied to the Preservation and
Restoration of Health of Body and Mind. By O. S. Fowler. 1s.

MEMORY AND INTELLECTUAL IMPROVEMENT; applied to Self-Educa-
tion. By O. S. Fowler. 6d.

HEREDITARY DESCENT; its Laws and Facts applied to Human Improvement.
By O. S. Fowler. 1s.

FAMILIAR LESSONS ON PHYSIOLOGY. Designed to aid Parents, Guar-
dians, and Teachers in Education. By Mrs. L. N. Fowler. 3d.

LESSONS ON PHRENOLOGY. Designed for the Use of Children and Youth.
By Mrs. L. N. Fowler. 6d.

INTEMPERANCE AND TIGHT-LACING; founded on the Laws of Life, as
developed by Phrenology and Physiology. By O. S. Fowler. 3d.

TOBACCO: Its History, Nature, and Effects on the Body and Mind. By Joel
Shew, M.D. 3d.

Vol. 1, containing the above, neatly bound in Cloth, Five Shillings.

THE NATURAL LAWS OF MAN. By J. G. Spurzheim, M.D. 6d.

MARRIAGE: Its History and Ceremonies. By L. N. Fowler. 6d.

FAMILIAR LESSONS ON ASTRONOMY. By Mrs. L. N. Fowler. 6d.

SELF-CULTURE AND PERFECTION OF CHARACTER. Including the
Management of Youth. By O. S. Fowler. 1s.

MARRIAGE AND PARENTAGE; or, the Reproductive Element in Man as a
means to his elevation and happiness. By H. C. Wright. 1s.

TEA AND COFFEE: their influence on Body and Mind. By Dr. William
Alcott. 3d.

EDUCATION: its Elementary Principles, founded on the Nature of Man. By
J. G. Spurzheim, M.D. 1s.

MATERNITY; or, the Bearing and Nursing of Children. Including Female
Education and Beauty. By O. S. Fowler. 1s.

Vol. 2, containing the last 8 Works, Cloth, neat, Six Shillings.

www.ingramcontent.com/pod-product-compliance
Lightning Source LLC
Chambersburg PA
CBHW021805190326
41518CB00007B/459